2

THE MATILDA MISSION

by BRIAN EDWARDS

COOLINDAH

Published in Australia by Coolindah
An imprint of the Weldon International Group of Companies
43-45 Victoria Street McMahons Point NSW 2060 Australia

First published 1993

Publishing Director: Leonie Weldon
Editor: Avril Janks
Production Manager: Cathy Wadling
Designed and typeset by Astri Baker
Printed by Australian Print Group

National Library of Australia Cataloguing-in-Publication data
Edwards, Brian (Brian Clifford)
The matilda mission.

ISBN 1 86302 279 1.

1. Edwards, Brian (Brian Clifford) — Journeys. 2. Flights around the world. 3. Tiger Moth (Training plane). I. Title.

910.41

FOREWORD

From the opening chapter, which tells about the disastrous start to this epic adventure, there is incident after incident that bears witness to the skill, determination and sheer flying ability of a man and the fulfilment of a dream.

The Matilda Mission is an account of one man's triumph against fearful odds and should be read by all, especially by every young Australian. It shows that the spirit that drove the early aviators is still with us, and, it is to be hoped, latent in the younger generation and only waiting to be released.

I would be surprised if any reader were not moved by this saga. It should go down in the history of aviation as an epic achievement.

Older pilots will, I am sure, be filled with admiration for and perhaps a little envy of the amazing exploits of Brian Edwards on the Matilda Mission.

Sir Fred Chaney, KBE, AFC

Sir Fred Chaney at the launch of the Matilda Mission, December 1989.

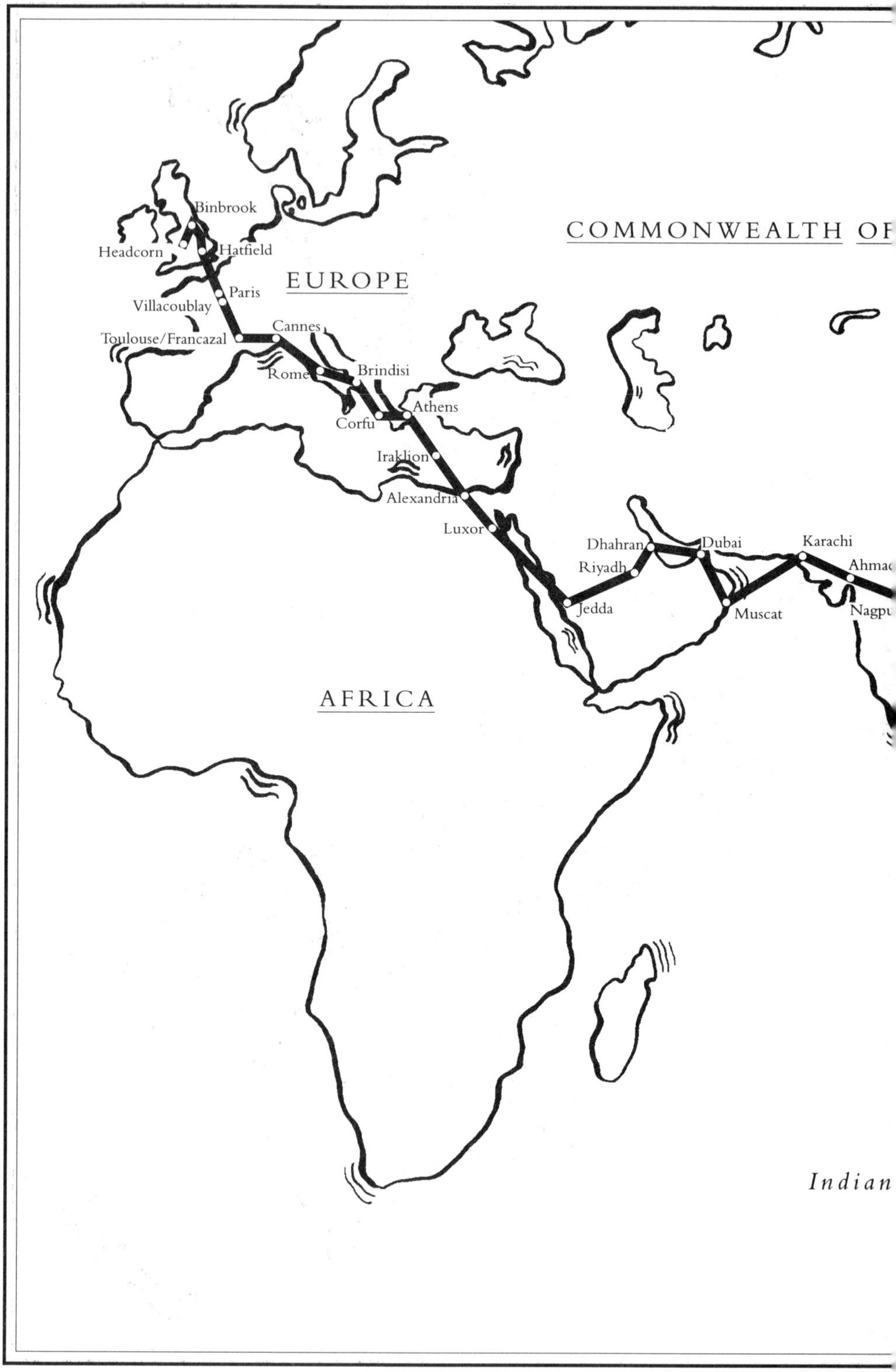
COMMONWEALTH OF
EUROPE
Binbrook
Headcorn
Hatfield
Paris
Villacoublay
Toulouse/Francazal
Cannes
Rome
Brindisi
Corfu
Athens
Iraklion
Alexandria
Luxor
Dhahran
Riyadh
Dubai
Karachi
Jedda
Muscat
AFRICA
Indian

ROUTE OF THE

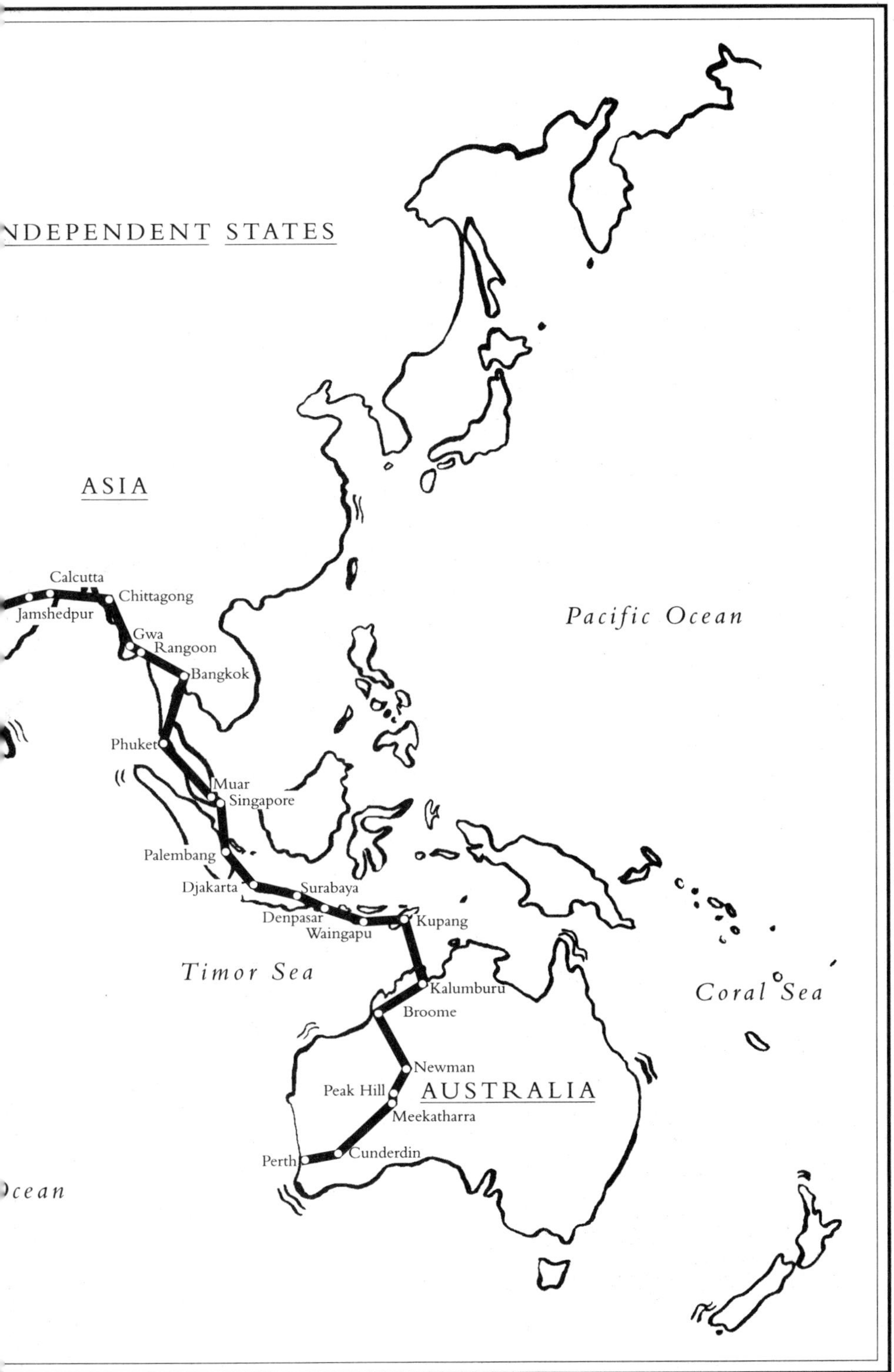

MATILDA MISSION

The Matilda Mission Committee.
Back row, from left: Kevin Somes, Tim Keenan, Tony Wilkins (Engineer), Tim Flynn, Sir Fred Chaney (Patron). Front row, from left: Rob Devenish, Michael Beech, Brian Edwards, Neil Donaldson, Peter Anderson (Chairman).

CONTENTS

Chapter 1

False Start—From Binbrook to Headcorn

'Mayday, Mayday, Mayday. Hotel Papa Hotel, Hotel Papa Hotel, Hotel Papa Hotel. I am at 2000 feet over Smarden, Kent, tracking 173 degrees—engine failure—will be carrying out forced landing.'

'Hotel Papa Hotel, this is Kent Radar, say again your position.'

'Hotel Papa Hotel at 2000 feet over Smarden, tracking 173 degrees.'

My God, this is a real one. I don't really want to talk any more on the radio. Before I couldn't pick up their radar. No problem now. I wish they'd stop asking questions.

It was March 2 1990. I had been cruising comfortably towards my refuelling stop at Lydd, a small airfield on the coast; the sun was out and the circulation was coming back into my hands and feet. My navigation so far had been good. I could see the English Channel and was on my way to France. The years of planning had finally come to fruition and I was beginning to feel comfortable.

Leaving on the first leg of the Matilda Mission—for the first time! Binbrook, March 2 1990.

When it came, the bang was horrendous, the shudder awful as it went through my small biplane. The control column slammed hard to the side against my knee as Matilda began to stall and go into a spin. The engine revved as though it would fall from its mountings. I quickly corrected—closed the throttle opposite rudder, control column forward, bringing her out of the stall, and levelled out, but was still unsure exactly what had happened. No oil pressure, a gaping hole in the top starboard wing and the fabric peeling off.

At first I thought I must have hit something, but no. The crankshaft had snapped. The propeller, flying off, had smashed through the upper starboard wing, splitting the leading edge.

A momentary feeling of being in a dream—a nightmare—came over me. This couldn't be happening. It was my first day. I could see the Channel. I was on my way to France, the beginning of my adventure.

Reality hit me. This was no dream—the dropping air speed indicator told me it was all too real. I had lost height and realised I had little time left to get us both safely on the ground. Thank God for past training. Quickly I moved into procedures for forced landing. The automatic responses of adopting optimum gliding attitude and looking for an appropriate field were quickly performed. There appeared to be plenty of paddocks available, and I picked out one that looked firm and flat with no power lines in the vicinity.

There was a large clump of trees next to a farmhouse over which I would make my final turn. I turned a little higher than 500 feet, making sure I had plenty of height to reach the field. As I turned I glanced at the trees, noting that the wind had changed. Matilda seemed to be handling all right, but the hole in the wing looked bad. I decided not to unhook the slats as the hole appeared to go right through the leading edge. I thought the slats might be damaged as well.

Turning over the top of the trees and keeping the speed at 60 knots, I gently dropped to 55 knots, and lined her up. I lifted the nose up a little to let the speed drop off.

The wind was on the nose. I turned everything off, and began to let her down gently. A hedge ahead, and I lifted the nose up a little. Speed was down to 50 knots. It was then that I saw the ditch that was hidden behind the hedge.

The starboard wheel slammed into the top of the ditch and I came to a sudden, jolting halt. I was thrown forward in my straps and silently acknowledged their effectiveness as my nose brushed the instrument panel. I was then thrown back into my seat as the tail fell backwards into the deep culvert.

FALSE START

I sat in the plane. I was down. Everything was very quiet. The wind wasn't all that strong—I could hear it, though. I thought, 'It's actually blowing through the flying wires. Funny, but I'm not in the air! I think I'm dreaming. Am I? No, I'm not. What about everyone at Binbrook? What about everyone at home? I've come down on my first leg! Fifteen years and this. Without even leaving the UK! I've let down a lot of people.'

Then a voice came: 'Excuse me, sir—would you like a cup of tea?' I glanced over my shoulder. Looking at me over the hedge was a ruddy-faced, beanie-clad farmer wearing a duffle coat. Would I like a cup of tea? Well, I suppose I would!

Slowly I undid my harness, unhooked my radio leads, took off my goggles and helmet, and hauled myself stiffly out of the cockpit. I shouldn't have stayed there anyway—not the best thing to do after a forced landing with some 240 litres of fuel still on board.

I stepped onto the starboard wing, which was now resting on the ground. Poor Matilda. The wheel strut had dug into the soft mud and she was lying at an awkward angle. I slowly walked around the plane. There didn't seem to be a lot of damage, but for the gaping hole in the wing where the propeller had gone through. It looked awful. The nose cowling was ripped open, another legacy of the freewheeling propeller, and the ugly jagged end of the crankshaft jutted out from the dented nose cowling. What a mess!

'Let's go and have that cup of tea,' I said to my patient farmer friend.

As we started to walk across the field, I heard the unmistakable sound of a helicopter. 'I hope it's not going to land,' was my immediate thought. There was obviously nothing wrong with me—at least physically—and the last thing I wanted at that stage was any interview. After all, I was going to have a cup of tea with my farmer friend!

I glanced around and saw that the helicopter had landed. There was a young lady climbing out of it, wearing a white uniform and carrying all sorts of coiled rubber tubes. 'Oh hell, she's a nurse,' I thought. It reminded me of something from M.A.S.H.

I endeavoured to quicken my pace, stumbling slightly over the ploughed ground and leaving my farmer friend behind. The zealous nursing sister, however, was fast closing the gap. She finally caught up with me, and took me by the elbow. 'Look,' I said, 'I'm perfectly all right—thanks for coming.'

'Oh, I can see that,' she replied in a controlled manner, showing all of her past training, 'but just come with me—you're in shock.'

'Shock!' I replied. 'There's no bloody doubt about that!'

The startled look on her face was accompanied by a slight step backwards. 'What do you mean?' she said.

'I'm in shock,' I replied, 'but I don't think you can help.'

'Well, I think I can,' was her retort. 'I've seen them like this before,' she said, turning to the farmer, who was now looking quite bewildered.

'You may have seen them like this before, young lady, but I'm not going with you. Please go back to your helicopter; thank you for coming.'

The rest of the conversation I cannot recall, but I do remember that after a few minutes she went back to her helicopter and I went off to have my cup of tea.

We tramped across the muddy field to the farmer's cottage on the edge of the field. It was a small neat brick house with a mottled-tile roof. I had noticed the unusual roof tiles as I turned over the top, coming in for my forced landing. We walked around the back to the kitchen door. My farmer friend went inside, but I stood at the door, rather reluctant to enter with such muddy boots. The farmer's wife beckoned me in, a small-framed, fine-featured lady, with her hair pulled tightly back in a bun. She was wearing a plaid dress and, as I recall, a red woollen cardigan. Her husband introduced me as 'the pilot who just came down in our field'.

She was surprisingly unruffled, but I really wonder what she thought. Here was her husband with a rather untidy aviator, with flying jacket a little torn, and covered in mud. What an extraordinary sight I must have been. As I extended my hand to her, I dropped my helmet and goggles onto the lino-covered floor. She took my hand, bent down, picked them up, and, handing them to me, said, 'Nice to meet you, but look after these, you'll probably be needing them.' She then turned to the stove and busied herself with the teapot and cups.

I leaned on the sink and looked out of the window. I could see my beloved yellow Matilda out in the field. She looked awfully alone and deserted, and the wind had picked up. The tea was scalding hot, and after one or two sips I poured it out. 'I really have to get back to my plane—I'm sorry. Thanks for the tea.'

I walked back across the paddock to the plane. Poor Matilda, my wounded Tiger Moth. She was lying at such an odd angle—a bird that had been shot down. I wondered if she would ever fly again. The engine was obviously finished. Although there didn't appear to be a lot of damage to the fuselage, the gaping hole in the wing looked worse. I rested my head on the cowling. 'What an absolute mess I've made of things,' I thought.

I lifted my head, to see a policeman walking across the field. On seeing me, and obviously having already given appropriate consideration to my predicament, he introduced himself by saying, 'Would you like to come to the station and have a cup of tea?'

Soon I was sitting across the desk from the superintendent in his small office. It was warm, but I was cold. I tried to drink the cup of tea and couldn't. The whole ghastly nightmare was real. I had crashed. The years of planning had come to nothing. I had let so many people down.

I put my head in my hands, and wept.

Chapter 2

On My Way

It had been some fifteen years in preparation: a solo flight from England to Australia in a Tiger Moth. There were three reasons why I wanted to make the flight.

Legacy, the Australian charity that cares for the dependants of deceased servicemen, cared for my mother and myself for twenty years. There was a debt to repay. I also wanted to fly from the UK to Australia, the Kangaroo Route, solo in a biplane, as Bert Hinkler and Sir Charles Kingsford Smith had done—what an adventure that would be! And I wanted to honour the memory of my father, Pilot Officer Clifford Edwards, pilot of a Lancaster Bomber of 460 Squadron based at Binbrook, Lincolnshire. He and his crew failed to return from operations over Cologne on July 3 1943.

The extract from my father's Squadron's Operations Record Book reads:

> 3/7/43 Weather—cloudy over the wolds and near the coast, fine inland becoming fine later. During the day several aircraft did their air tests. During the early part of the evening, prior to going on operations, two aircraft which were bombed up ready to leave caught fire and were completely destroyed and three others became unserviceable. For a space of half an hour or so there was much excitement as the aircraft were loaded with 4000-pound bombs, which fortunately were not fixed. After several terrific explosions which caused not a little damage, obliterating several huts in the vicinity and smashing hangar windows, the fires were extinguished and over an area of about 500 yards pieces of the aircraft were scattered. Fortunately no one was injured although several of the ground staff were in the aircraft when she first caught fire and had to run for it.
>
> Seventeen aircraft eventually left and attacked Cologne. This operation was considered a great success as many large concentrated

> fires were left burning fiercely. From this operation, unfortunately, one aircraft failed to return. The aircraft crew were as follows:
> F/SGT EDWARDS C, PILOT AUS 421093
> F/O SPIER WF, B/A AUS 22378
> F/SGT TRUSCOTT CF5, W/OP A/6
> 1074494 SGT YOUNG C, F/ENG
> 1623946 SGT TYTHERLEIGH
> AE A/G 1317610 SGT REES, WJ
> A/G AUS 420653 SGT FRAZER LJ
> This aircraft took off from base at the appointed time, 23.29. Since then, nothing has been heard of either the aircraft or the crew.

Once I decided on my flight, the first requirement was to find out what flying in a Tiger Moth was like, as I had done all my flying until that time in a covered-cockpit aircraft. I called the Royal Aero Club at Perth's Jandakot airport and made a booking. Fortunately the Tiger Moth was free that day, as was an instructor.

I still remember well walking out to the plane and being shown how to carry out the daily inspection: testing the flying wires for tension, checking the split pins in the aileron and elevator, checking the cables for movement and no wear. Really quite simple, but effective.

Climbing into the rear cockpit and putting on my harness and helmet were not quite as simple as they sounded. For a start, getting in was awkward. I didn't know what to hold on to or where to put my feet, frightened, of course, that I'd damage the fabric. Once settled in the plane, harness and helmet on, I found I was far too low and couldn't see as much as I would have liked. I had always thought sitting in the open air would mean being able to see everything clearly, but here I was somewhat 'snuggled down' into the cockpit.

To alter this was quite a performance. It meant undoing the harness, taking off the helmet and goggles, getting out of the plane, walking around to the locker, taking out the cushions and placing them in the seat, and then climbing back in and putting harness and goggles back on. It took me ten minutes!

The simplicity of the instrument panel took me by surprise: altimeter, tachometer, turn and bank indicator, oil pressure gauge and air speed indicator; on my left-hand side a throttle, and on my right a slat lock. I smelt the faint aroma of well-used leather, however, and that made me feel more comfortable.

The instructor ran quickly through the start-up procedure: 'Switches off, fuel on, column hard back, pull the propeller through four times ...

Switches on, throttle set, column hard back.' And, as this Tiger Moth was fitted with brakes and a tail wheel, 'Brakes on, contact.' One pull, and the engine coughed into a comfortable rhythm.

On taxiing out to the runway I was told, 'You will find that the use of the rudder and ailerons on the ground are far more critical with the Tiger than with the usual tricycle undercarriage.' This, I was to find, was something of an understatement.

Lining up at the end of the runway, in a gentle five-knot crosswind, I was given the instruction, 'Clear for take-off,' from Air Traffic Control. (I will refer to this as ATC from now on.)

'Put on full power smoothly and move the control column forward, lift off the tail wheel—not too far forward, or the propeller will chop onto the runway,' said the instructor. It was a smooth take-off, and we climbed to 500 feet, made a gentle climbing left-hand turn up to 1000 feet, reduced the power and made a gentle left turn onto downwind.

'Tiger 37 downwind,' I called.

'Tiger 37, follow the Baron late downwind,' replied ATC. I turned onto base and reduced power, trimmed her back to 65 knots, and turned onto final at 600 feet, gently losing height.

'Clear to land,' came the instruction from ATC. Then the difficulties started. The small Tiger Moth jumped all over the place. I couldn't keep her under control. I made an over-correction of the rudder, elevator, aileron! This was going to be impossible. I began to get that awful feeling of clumsiness that happens from time to time when you realise things are not going well.

I looked up into the rear-vision mirror outside the front cockpit. The instructor had a broad grin on his face. 'Keep at it—she will really test you,' he said with a slight chuckle. I did keep at it, and some five hours later went solo.

The long haul to becoming a competent Tiger Moth pilot was not without frustrations. Without being over-confident, I considered myself to be a safe and reliable pilot and my experiences had taught me not to take things for granted and to plan methodically for long crosscountry flights. But Tiger Moth flying was a little different. Crosswind conditions up to 15 knots are manageable with the tricycle undercarriage. With the tail-dragger biplane, however, crosswinds above 10 knots required considerable practice. I took a break from this hard work to indulge in some simple aerobatics—loops and spins kept up my enthusiasm for eventually mastering the Tiger Moth.

Slowly, as my technique improved, so too did my confidence. I began to sense the full enjoyment of open cockpit flying—having the wind on my face, and looking across the wings and seeing the slight vibration in the flying wires, the wooden struts and the simple fuel indicator between the top wings. Holding the control column between finger and thumb was sufficient to direct the sensitive little biplane.

Although there was a long way to go, the excitement of the trip—in the footsteps of our great aviators Bert Hinkler and Sir Charles Kingsford Smith and flying the Kangaroo Route as they did, in an open-cockpit biplane—would occasionally engulf me and fill me with a sense of anticipation. During this period I thought a lot about my father, my mother, Legacy. I'm not sure why. It wasn't as though these things had been constantly on my mind in the past. At 49 years of age, with my own family, I had little time to dwell on my past life, but the father I had never known and the debt to Legacy, which had looked after my mother and myself in my first twenty years, began to be part of the whole structure of the pending adventure.

The following months saw me flying as much as possible, and starting to give some thought to obtaining my own Tiger Moth. This proved to be quite difficult, but negotiations were commenced on a contract to purchase a rebuilt Tiger Moth in New South Wales. This rebuild was destined to be Matilda (she had no name then). A 49-year-old man buying a 47-year-old plane for a trip of 12 000 nautical miles. Was I mad?

Momentum then began to slow; I was not in a position to commit myself financially to such an undertaking, since I had to buy two years after agreement to purchase. The usual pressures of home life, teenage children and work began to consume my time, and I found it increasingly difficult to maintain my commitment to the project.

A year had passed; I was not flying as much. I was no closer to committing myself to the rebuild project. It looked as though my dream was destined never to be translated into reality.

One Sunday morning I walked into my study and slumped in my chair. I reached up to the bookshelf and pulled out an old school atlas. I turned to the first page, a map of the world. I started to look at the route from England to Australia and the countries between, the countries I would fly through. It was time to pick up the threads and continue with the planning.

About that time there was a small article in the paper concerning an Englishman, Flight Lieutenant David Cyster, RAF, who had completed a flight from the UK to Australia in a Tiger Moth in commemoration of Bert Hinkler. I immediately wrote to him, explaining that I was planning a similar flight and asking if he could give me any assistance.

I received a very prompt reply from him, giving me details of his route and some of the modifications he made to his plane. He also wished me all the very best with my project.

When I mentioned to my wife Jenny that I had received a reply, she surprised me by saying that it was about time I got on with it. This, coupled with a question from my eldest son, Michael—'When are you doing your flight, Dad?'—was the catalyst that put me back to planning the flight again.

Operational navigational charts were difficult to obtain, but after many letters and phone calls over a period of some three months I finally obtained a set from the Aircraft Owners and Pilots Association in the USA. There were 23 charts in all, large and bulky, and it was obvious that once I had decided on the exact route I would have to cut them down quite considerably, to be able to hold each map in one hand.

I had decided that I did not want any leg to exceed 520 nautical miles. This would mean a time in the air of between six and eight hours, which really is quite long enough in a Tiger Moth. My route would not be the same as Bert Hinkler's, as travel through some Middle East countries was impossible, but it would be close to it. I planned to depart from Binbrook, Lincolnshire, my father's old base.

With the route roughly planned, I went back to the problem of acquiring a Tiger Moth. This meant further negotiations, and finally a commitment from me. A rebuilt plane was arranged, including a fully reconditioned Series I Gypsy Major engine.

Over the next few months there were constant minor alterations to the route because of political uncertainties in various countries. Fortunately, I was able to have the assistance of Jeppeson Memrykord, an England-based international flight planning group. Without their help the finalising of many details would have been impossible.

Aerodromes of entry, visa applications, availability of fuel and oil, and ground handling all had to be arranged, as well as specific time schedules. Countless phone calls to the UK during this period were remembered only by the size of the telephone account at the end of each quarter. But slowly the details were becoming finalised.

Simple hops through Turkey and the Middle East appeared unlikely. Relatively long water crossings over the Mediterranean, Red and Arabian seas and an awkward leg over mountains from Rangoon to Bangkok appeared to be highly probable.

The Tiger Moth took a little longer to be completed than originally anticipated. However, I eventually took delivery in September 1988, and flew her back to Jandakot, Western Australia from New South Wales. This was my first long crosscountry flight in a Tiger Moth, and planning it was good experience for me. My frequent stops across the Nullarbor provided me with invaluable experience in both dead-reckoning navigation and handling difficult short field landings in strong crosswinds.

When flying a Tiger Moth you have both feet on the rudder pedals, the left hand on the throttle and the right hand on the control column. There is no autopilot, so constant control has to be maintained. The course map has to be held in the left hand as well. The 70-knot wind that blows over the cockpit makes the latter task quite interesting!

I picked up a friend of mine, Rob Devenish, at Port Pirie, and he came along as passenger and assisted with the navigation. It really was a lot of fun landing on bush strips, and taxiing up to the petrol bowser for fuel, to the gasps of disbelief from onlooking motorists. There were occasional awkward crosswind landings, and I was beginning to understand the difficulties of dead-reckoning navigation over these desolate and remote areas with only compass and map. This, however, was combined with the quite extraordinary thrill of slowly mastering the Tiger, flying low over roads and small towns, trying to keep up with the road trains, and becoming competent at holding a map in one hand and the control column with the other.

My first water crossing was experienced on this trip: nine nautical miles over St Vincent's Gulf. I was a little nervous when I took off from Port Pirie, but settled down quite quickly when I thought of the very long crossings I was soon to undertake.

The arrival of my Tiger Moth at Jandakot in September 1988 was significant in the sense that the next step would now have to be taken. Modifications to the aircraft and finalisation of the route were indications that my flight was now a serious project, and was going to happen.

Although I had not said much about my proposed adventure to many, those to whom I had mentioned it usually just smiled and shook their heads. I think it was viewed with a mixture of disbelief and the conviction that I was either daydreaming or mildly stupid.

My family, of course, were very supportive. Michael, 20, Christopher, 19, and Susie, 17, all kept asking about the departure date, and Jenny was enthusiastic (but, I suspect, secretly hoping that it wouldn't happen).

The modification program began in earnest. The Tiger Moth has a fuel capacity for two and a half hours. This had to be increased to around ten hours. As well as this, extra oil would be needed. Graham Swannell of Aeronautical Engineers of Australia designed a series of fuel tanks for the front cockpit, from which the seat, instrument panel and control column had to be removed to accommodate them.

The system that was finally decided on was a series of three tanks, one of which was the usual 45-litre auxiliary tank, and a further two ferry tanks, one of 150 litres and the other of 35 litres. The fuel from these tanks had to be pumped up to the top main tank between the two upper wings. It would then gravity-feed down to the engine.

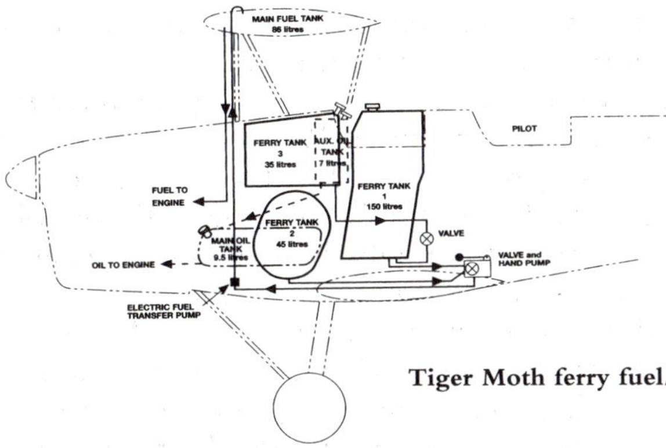

Tiger Moth ferry fuel/oil tanks.

The modifications, plus the usual 85-litre tank between the top wings, would give me a total fuel capacity of 315 litres. At a fuel consumption rate of 31 litres per hour, it meant I had ten hours of flying time available. At a true air speed of 70 knots, I could travel a maximum of 700 nautical miles in still air.

The fuel indicator in a Tiger Moth is a simple float in a glass sight tube situated on the top of the main tank. This is clearly visible from the rear cockpit. To augment this I had a sight tube fitted to the fuel pipe that transferred the fuel from the auxiliary and ferry tanks. When one tank was empty, air bubbles would be visible in the sight tube, and I would then change to another tank. I had an electric pump and a hand pump fitted.

The Tiger Moth uses about one litre of oil every hour, and the standard small oil tank did not contain enough for ten hours so a further auxiliary oil tank was placed in the front cockpit. The oil gravity-fed to the main oil tank on the port side of the fuselage. The pipe was transparent and acted as a sight tube, which I could see by looking out over the side of the cockpit. The front cockpit was then enclosed with a sheet of plywood.

Other modifications included a wind-driven generator slung below the fuselage, a VHF radio, a transponder and an Automatic Direction Finder (ADF). I installed another compass just above the crash pad. The original compass was losing its great efficiency, being over 40 years old now: it was starting to become difficult to swing accurately. I was to be grateful for this precaution later.

Many weeks were spent in the Royal Aero Club hangar completing these modifications. Tony Wilkins, an engineer with the Club, undertook most of the work, in particular being assigned to the job of remodelling the rear cockpit to allow for the extra equipment. Slowly things began to take shape, although there was no doubt the cockpit would be somewhat cramped.

Testing of the fuel system took a little longer than expected, but eventually it worked satisfactorily. I was now confident that the system we had was reliable and efficient.

I chose a departure date of March 2 1990, and, with much help from Jeppeson Memrykord's Bob Catt, who had undertaken to handle clearance details, arranged with the Royal Air Force to take off from RAF Binbrook in Lincolnshire—my father's old base. I named my Tiger 'Matilda'. After all, I had always thought Waltzing Matilda should be our national anthem!

A committee was formed, of close friends: Michael Beech, Kevin Somes, Tim Keenan, Tim Flynn, Neil Donaldson, Rob Devenish, and Peter Anderson, Chairman. They worked enthusiastically for the 'Matilda Mission', as it had now become known. Len Findlay, a feature writer with the West Australian, our major newspaper, wrote a superb article to launch the project and supported us greatly. Neil Donaldson produced a beautiful brochure, and we all felt confident sponsors would be found.

But the somewhat arduous task of seeking sponors was made more difficult by the failing Australian economy at the time, and a large corporate sponsor did not eventuate. The project was supported by many friends, however, and small business groups, and, with the help of Channel 7 and the West Australian, a significant sum was raised from donations or pledges.

Perth Legacy had become an integral part of the preparations, giving their full support. The members of my father's old Squadron 460 joined in the adventure, making me an honorary member.

Launch Day was Sunday December 10 1989. The Civil Aviation Authority had given me permission to land at Langley Park, an extremely picturesque strip of parkland separating Perth city from the Swan River.

In the early 1930s Langley Park was Western Australia's first commercial airfield, being used by Sir Norman Brearley and Sir Charles Kingsford Smith. Nowadays it is used by the public, and on most Saturdays and Sundays is completely covered with people playing all forms of sports, their colourful jumpers and tunics making a dazzling display.

On this Sunday Langley Park seemed different to me. It was a clear, bright morning, and as I flew down the Swan River over the Narrows Bridge accompanied by the Royal Aero Club's Tiger, piloted by John Douglas, the Club's Chief Flying Instructor, I knew I was about to land on hallowed ground.

The Edwards family at the Langley Park launch, December 1989.
From left: Christopher, Brian, Jenny, Susie and Michael.

As we turned onto final I could see a small group of people clustered around a blue-striped marquee at the end of the makeshift runway. I was a little nervous, as it was my first landing at Langley Park, and, being aware of the historical significance of the Park as far as Western Australian aviation was concerned, I was conscious of the need to make it a perfect one.

When I taxied up to the marquee a light patter of applause broke out, and as I cut the engine the small group of supporters milled around the plane.

Sir Fred Chaney, who had agreed to be the patron and had been a Tiger Moth instructor in World War II, officially launched Mission Matilda. It was an enjoyable morning, and an amount of some $40 000 was pledged in half an hour. My committee were ecstatic: they had all been working very hard, and had come up with the idea of selling each leg of the flight. It was a brilliant move, and it enabled many people to be intimately involved with the flight. My old school, Wesley College, was one of the first to offer support. Many people were beginning to get involved, and although I was very grateful for their support I started to realise what responsibility I now carried.

My family had always accepted the fact that I had to attempt the flight, but now many other people were also expecting me to succeed. I felt a slight twinge of apprehension. I had not thought that this would be the case—I had always thought that it would be just 'me and Matilda'. But now it would be much more.

As I took off to fly back to Jandakot, I briefly reflected on what it would be like on the day of my return. Would there be people to welcome me home? Probably not. I quickly put it out of my mind. As there was so much yet to be done I did not want to allow myself the luxury of such thoughts. Taking one day at a time was very important; I was going to learn this as the Matilda Mission unfolded.

Over the next two weeks Matilda was dismantled, wings, tail-plane and undercarriage being removed and secured in a container and shipped off to England in early January. Tony Wilkins and I left three weeks later.

The Matilda Mission had truly begun.

Flying over Perth before leaving for England.

Chapter 3

England

I

Tony and I arrived at Heathrow at 7.30 am on Friday February 4, after a pleasant flight with British Airways. My flight back would not be as comfortable! It was a typically cold, wet and windy English February morning, but we were cheered to be met by my Aunty, Gwen Gear, my father's only sister, and her son Duncan. Gwen is a small lady whose appearance and manner belied her then 78 years, and Duncan a tall grey-haired man with an accent and complexion that left no doubt as to his origins.

By an extraordinary set of circumstances, Bob Catt at Jeppeson Memrykord had arranged with the RAF—whose marvellous assistance was gratefully appreciated during the following three weeks—to have Matilda hangared at RAE Bedford, a Ministry of Defence airfield some 50 miles north of London. Gwen and Duncan happened to live in Bedford, and generously offered to put us up. I would stay with my aunt and Tony at Duncan's charming hotel, Woodlands Manor. We were 10 minutes from the airfield. It could not have been more convenient.

I had set a departure date of March 2 from RAF Binbrook, and this would give us just a little more than three weeks to assemble Matilda and test-fly her. It was going to be a tight schedule, but I didn't want to waste any time.

I had begun to feel that my impending trip would mean a lot more to many people than I had realised. To me the trip was being made in memory of my father, as a fund-raiser and as an adventure. But it was strange how it was taking on more meaning. Many people were now involved, enthusiastically supportive. For many I was doing something that they wished they could do. The planned flight was bringing back memories, long forgotten. Old and young alike found something to identify with. The sense of responsibility began to weigh heavily on me.

Early on the Monday morning Tony and I arrived at the gates of RAE Bedford and were met by Wing Commander Vic Lockwood, the Base Commander, and Squadron Leader Harry Shaw. The former was the typical young, polite, decisive RAF officer: 'Delighted to meet you, Brian—hope you find everything OK. Looking forward to seeing that Tiger Moth of yours. Are you sure you know what you're doing?'—this last said with a good-natured chuckle.

Harry Shaw was a small, neatly dressed man who wore a quite superb small hat, as I recall. The wind, which was very strong that day, appeared to pass him by without ruffling either his coat or the hat, whereas the rest of us were quite dishevelled by the very same wind.

They made us very welcome at RAE Bedford and provided hangar space—Hangar 3, to be precise—to reassemble Matilda, and the personnel were all extremely helpful. On many occasions Tony and I would require special nuts, bolts and such like; all were cheerfully provided—nothing was too much trouble.

The next three weeks were a mixture of frantic work on Matilda, desperation while trying to finalise departure procedures, and occasional despair and frustration at the difficulty of obtaining visas for some countries. The latter, I had been assured, would be easier to arrange in the UK than in Australia, but of course this proved to be entirely wrong.

Value Added Tax (the British equivalent of Australia's proposed goods and services tax) was at first a problem for us: we had a request for a cheque of some 10 000 pounds sterling by the British Internal Revenue Department. But this was very expertly handled by our Agent General, Mr Ron Davies. I gather there was no VAT payable on vintage cars. Consequently, Matilda became a vintage car for 48 hours!

Tony and I worked at a furious pace, and in two weeks had Matilda ready to test-fly. The weather had been appalling, with gale-force winds and storms. The whole of England was still gripped by the icy fingers of winter. Massive lows came constantly across the UK, bringing the dreadful conditions, and no abatement was in sight. Frustrations piled up. Test-flying was extremely important, but in such conditions it was impossible. Although everything had worked satisfactorily in Australia, Matilda's long trip in the container and then the reassembling gave plenty of opportunity for the gremlins to establish themselves.

During this period, Harry Shaw and I spent much time going over the route, particularly the European legs. Here Harry's local knowledge was invaluable. The map for each leg and all the relevant information, aerodrome diagrams and procedures were placed in separate envelopes marked on the outside with the destination—'Binbrook–Villacoublay' for

Day One, up to 'Kupang–Australia' for Day 23. I doubted I would make it in 23 days, but that was the goal. After all, Hinkler had made it in 15!

Squadron Leader Roger Kissane at RAF Northholt (Royal Flight and Special Briefings Branch) had provided me with a complete set of low-level charts for the route, and the appropriate information from these was transcribed onto my Operational Navigational Charts. Although I had a small pouch next to my seat for other maps, changing maps over in the cockpit was often difficult. The terrifying prospect of having the map blown out of one's hands was always present, and a very firm grip was necessary. Coupled with this, one had to be dextrous enough to change the folds in the map with one hand while flying from one point to the next. It took a little practice, but it was important to have the map folded, with all of the appropriate information on it, in such a way as to make the task as simple as possible.

Although I had an Automatic Direction Finder (ADF) I had decided that I would not rely on this, and that my visual navigation was to take precedence over everything else. Apart from the fact that there were few non-directional beacons in Europe and the Middle East, I had been told that on many occasions they malfunctioned or were poorly serviced. As well as this, the ADF is often affected by the weather (it points very accurately to thunder storms!) and terrain. Compass headings, track, drift, visual fix and ground speed were to be my essentials for each day's navigation. The weather was something I would just have to cope with.

The weather finally cleared sufficiently for me to test-fly Matilda on 22 February.

I felt we had taken far too long to get her back together. I'm sure I pushed Tony from time to time, but I felt neither he nor anyone else really understood the urgency of the matter. The Matilda Mission was now consuming me.

The day before my first test flight Harry came across to the 'office' that Vic Lockwood had made available to us, a small, disused wooden-walled room that still had the smell of 'Flying Operations'. I suppose this is an unusual comment to make, but to me it was a wonderful feeling, thinking that I had been taken into the 'bosom', at least as far as I was concerned, of the RAF. This gave me a sense of warmth, a feeling that maybe they understood what I was trying to do. This is something that was only now beginning to be obvious to me. All the doubts, the failures, the despair, the struggle—both physical and mental—that had dogged my existence for some 40 years might now in some odd way be laid to rest.

'Tomorrow is fine for you to test-fly, Brian; here are the aerodrome procedures,' said Harry Shaw, handing me a slip of paper. 'We summarised them for you. The aerodrome will be fairly quiet tomorrow, so you will have the runway and airspace to yourself!'

We were now in the final stages, and that night I proudly phoned Jenny at home in Perth to tell her that I would be test-flying the next day. I went to bed early, but lay awake looking out of the window at the black, cloud-filled sky. I still felt a little nervous about the trip. I knew deep down that I lacked the flying experience to undertake such a trip, and I had never flown outside of Australia. Doubt engulfed me. Oh God, what was I taking on!

After a somewhat restless night, I awoke and looked out of the curtained window of my room at the flag hanging limply on the pole at the Airforce Association's clubrooms across the road. Maybe this would be an easy day for Tiger Moth flying.

With the exception of the ADF, which was still occasionally inconsistent, the test-flying went off without a hitch. Everything appeared to go well—handling characteristics, fuel transfer and rates of climb. I had not lost my touch, and despite the fairly strong winds flew comfortably and confidently and completed all tests in two days.

The RAF at Bedford continued their support, to the extent that after my successful testing Vic Lockwood asked me if I would like to fly in formation with a Harrier. Would I ever! He suggested that I should meet with the Harrier pilot and discuss proceedings. We drove to the other side of the airfield—the operational section—and I was ushered in to meet Squadron Leader Peter Collins, an extraordinarily young man, I thought at the time, to be flying such an aircraft. But I suspect my interpretation of his age was a product of mine! We discussed how we might make rendezvous over the hangar at Bedford for a photo to be taken. He made constant statements about time intervals, height and rates of climb, all of which filled me with absolute dread. Me, an ordinary private pilot, talking to a sophisticated RAF fighter pilot about formation flying!

We talked for 30 minutes and I hurriedly tried to write down everything on my crumpled pad. Next day we would be taking to the air together.

The wind had freshened overnight and was strong and blustery next morning. We rendezvoused by radio some ten miles to the east of the airfield, myself at 3000 feet, Peter at 10 000 feet. I descended to 500 feet over the airfield and nervously tried to keep my speed up to 80 knots with a slightly nose-down altitude. Suddenly there was a shattering roar, and I glanced over my left shoulder to see the massive fighter slowly creeping up on my port side. It was an awesome sight, and I must admit I was quite frightened.

We made two passes across the field and then landed. I taxied over to where the Harrier was parked and was greeted by Peter: 'I don't know how the photo will turn out, but that was a lot of fun—you really don't go very fast, do you!' he quipped laughingly.

We chatted for a while about Tiger flying and finally he said, 'Well, all the best for your trip, Brian—it's quite a long way to fly in any aircraft, let alone a Tiger Moth. I'd like to give you something. Here, take this—I don't think it will be of much use, but you never know.'

With that, he handed me a small pocket knife on a chain. I slipped it into my flying-suit pocket and thanked him, and we shook hands. It had been a very good morning, and a wonderful flying experience for me.

Gwen had been simply marvellous about putting up with my early-morning departures, my late arrivals home and the untidiness in my room, which was always in a complete mess, with maps and flying gear strewn all over the floor. I think she felt secretly it was the 1940s over again; my father, her only brother, had stayed with her in his time in England during World War II. She often mentioned to me that I appeared to make as much mess as he had. I think she enjoyed the three weeks as much as I did.

LEFT: **Harrier pilot and RAF Squadron Leader Peter Collins.**

RIGHT: **Flying in formation with the Harrier.**
Courtesy of the Royal Aerospace Establishment MOD (PE), Bedford, England, and the Public Relations Office, RAE Farnborough, England.

A few days before I was due to depart, Jenny came over with Tim Flynn, our public relations organiser. I got to Heathrow early to meet them, and eagerly stood waiting in the arrival lounge. I finally saw Tim's bustling frame pushing through the crowd.

'G'day, Eddy—good flight—bad food—when do I start work?' were his opening remarks as he took my outstretched hand. Jenny looked fresh as always, despite the long flight.

'You look tired, darling—is Matilda ready for the long flight home?'

'Not quite—but almost.'

A few hours later we were back at Bedford. Jenny would stay with me at Gwen's and Tim with Tony at Woodlands Manor.

Jenny, in her quiet manner, managed as always to give me great support in these last few days, and Tim in his own inimitable way secured clearances for Myanmar and Saudi Arabia. I'm not sure how he managed this, but I still remember the day he came running into the hangar at Bedford waving a piece of paper above his head and shouting, 'I've got them both.'

This was a great relief, as I had made a number of trips to London and numerous phone calls to the embassies, all to no avail. 'Just how did you do it, Tim?' I asked.

'Well, I don't really know. At one embassy I just walked in, said I was Brian Edwards, and signed your name on the papers. The only comment was, "You seem to have put on weight in the last week, Mr Edwards!" Then they stamped your passport! Easy, really.'

Two days before I was due to leave Bedford, I had an early morning phone call: 'Wooldridge, Daily Mail, Brian—how are the preparations going? Mind if we come up and take a few pics?'

I arranged to meet him at the guardhouse at RAE Bedford the following morning. Harry and I stood at the boom gate, passes in hand, as Ian Wooldridge and his photographer drove up. They got out of the car and walked towards the gate, and, shaking the guard's hand, Ian said simply, 'Wooldridge—Daily Mail.' It seemed to be all that was necessary—passes were not required.

Unfortunately the wind was gale force that day, and I was unable to fly. The best we could do was push Matilda out on the apron. Even then it took six men to hold her steady in the wind. Back in the shelter of the hangar, Ian, obviously a little unnerved at the apparent vulnerability of Matilda, said, 'Er—Brian, long way to go in that sort of plane, you know.'

'Yes, I suppose it is,' was my reply.

The tension was enormous these last few days. Both Tony and I had been working at a furious pace to have Matilda ready, and I'm afraid I was becoming a little more difficult than usual to live with.

In some respects this came to a head a day before I was due to leave Bedford for Binbrook. Tim and I had spent some time discussing departure details. Tim, a PR man after all, wanted a little 'pomp and ceremony' on leaving Binbrook. I could think of nothing worse. I just wanted to get going.

Somewhat troubled after our meeting, I arrived home that evening feeling particularly irritable. I walked into the loungeroom and slumped into the chair, having tossed my flying kit into the corner of the room with a little more force than usual.

Jenny and Gwen were quietly going about dinner preparations leaving me to my own thoughts, assuming that in such a mood I was best left alone. A loud thumping on the front door interrupted my brooding, and Gwen, on opening the door, was confronted by a youth, who was complaining about someone having parked behind his car.

As I had driven in that night, I had noticed that Gwen's car space was, as usual, cheekily occupied by someone else. I usually took no notice, but on this occasion I had parked behind the offending car, preventing the intruder from leaving. The youth at the door was the owner of the car, and he was complaining about the inconvenience! This was the last straw!

I leapt to my feet and went to the door, to be confronted by a sallow, barefoot youth. What followed is not something of which I'm particularly proud. Suffice it to say that I chased him and his three mates around the car park, fortunately not being able to catch any of them. In my own way, and in a way they obviously understood, I indicated to them how I felt. We did not see them again. Maybe I overreacted, but I felt better afterwards.

On March 1, I took off from RAE Bedford for Binbrook via Scampton, as the Commanding Officer of Scampton, Group Captain Howard, had very kindly invited us to have lunch with him on the way to Binbrook.

The dreadful weather had eased a little, and I departed at 10 am local time. Although the cloud base was quite high that day, the wind of 25 knots was almost on the nose; the 65 nautical miles to Scampton in 80 minutes—a ground speed of 50 knots—still made a pleasant flight.

On landing at Scampton, I was met by the Group Captain, and Jenny, Tony and Tim, who had come up by car. We all enjoyed Scampton's hospitality at lunch, and a display by the Red Arrows—the RAF's superb aerobatic team.

While at Scampton, Tony and I continued our efforts to swing my compass. It was freezing, and out in the compass bay the forecast 25-knot wind had freshened and the chill factor had increased. The Group Captain mentioned that it looked as though it would snow.

After we had spent two hours out in the compass bay the compass swing was completed, and we taxied back to the hangar. Tony and I were frozen. Before I left, the Group Captain read a telegram from Air Marshall Sir Michael Graydon, Air Officer Commander-in-Chief, RAF, wishing me all the best. This really gave me quite a lift.

I departed for Binbrook at 5 pm local time, after a small snowstorm had hurried across the Scampton base. The 45-nautical-mile leg across to Binbrook was made in the last of the watery twilight, and I was escorted by one of the Scampton Bulldogs, an RAF training aircraft.

It was a beautiful flight across the flat, snow-covered countryside, the last, dull rays of the setting sun behind me. After a short time the large but now disused runways of Binbrook slowly drifted into view over Matilda's cowling. I was given clearance to land on runway 27.

Throttling back, I came around onto base and gently lost height and speed. I turned onto final; the runway threshold grew larger, and as I descended the last rays of the sun were slipping behind the hills in front of me. On touching down, I taxied over to the massive hangar the RAF had given me permission to use overnight. These hangars were the very ones used for the Lancasters of 460 Squadron during World War II. I wondered fleetingly if they hangared Pilot Officer Clifford Edwards' Lancaster R for Romeo on the night of July 3, 1943.

The huge hangar doors were slowly cranked open, and Matilda was pushed inside by the small group of well-wishers who had arrived for the sole purpose of giving me a warm send-off the following morning. The hangar appeared larger inside than it did from outside. My tiny biplane had been literally swallowed up, and as we closed the doors, having secured her snugly inside, I felt a slight shiver of apprehension run through me.

We spent the night at a small hotel, 'The Limes', in Market Raison. There we had a farewell dinner with the group of friends who had made the trip up from London: Aunty Gwen and Duncan, Glen Scott, my sister-in-law, and her husband Mike, my nephew Blair, his friend Jeremy Edwards and the Western Australian Agent General, Ron Davies. Ian Wooldridge from the Daily Mail, Friends Di and Peter Osbourne, Tony and Susan Bird, John and Marie Gregson and their grandson and of course Tim and Tony had also braved the cold weather to come and wave me off.

Ron Davies and Ian Wooldridge expressed good wishes on behalf of everyone present for a safe and pleasant trip. Jenny and I retired early to our room, and for the last time I went over my first two legs of Day One: Binbrook to Lydd, refuel, Lydd to Villacoublay.

The leg to Lydd would be 160 miles—simple and straightforward, I thought. I should have known better!

I awoke at 5 am. Everything had been prepared; maps and flight plans were ready. I felt comfortable and confident. I pulled the curtain back and looked out at a black, star-filled sky and the white, quilt-like covering of snow on the ground and trees. Six inches had fallen overnight.

As we drove out to the airfield it began to get light. When we reached Binbrook there was a wind of 30 knots blowing across the airfield, the runway was iced over, and the hangar doors were frozen shut. The next three hours were spent clearing the ice off the runway (the RAF once again came to my assistance, washing the massive hangar doors with de-icing fluid), getting weather reports from the tower and submitting my final flight plan. Although the wind was forecast at 35 knots up to 3000 feet, it was forecast to be down to 18 knots at Lydd, and it was almost on the tail. The sky was brilliantly clear and sparkling by 9 am, but the temperature was minus three degrees.

My performance charts indicated I would need about 350 metres to get off the ground, just the distance cleared for me by the RAF team.

Shortly before I had left Perth, John Leaversuch, the Chief Examiner with the Civil Aviation Authority, had rung to wish me good luck.

'You know, Brian, both Melita and I think it is a long way to fly, but Melita's main concern is that you will probably get fairly cold sitting out there in the open.'

The following morning Melita presented me with a beautifully knitted woollen scarf. 'This will keep you warm,' she said, and now here I was wrapping the scarf around my neck. I was certainly going to need it. Farewells over, I taxied out to the cleared runway, lined up, full power, stick forward. Matilda seemed a little sluggish, but the speed slowly increased and I lifted off just short of the cleared part of the runway, clipping a small mound of snow that puffed around my slowly lifting Tiger Moth. One low pass and a wing dip to my waving friends, then I set course for Sleaford and climbed to 1500 feet.

The snow-covered ground made navigating difficult, and at 1500 feet the temperature was even lower than on the ground, and the chill factor heaven knows what! It was so cold that my hands and feet were completely numb. My only indication that I was holding the map tightly was its crumpled state.

On reaching Sleaford, I was immediately directed off course for some twenty minutes, due to other traffic. I was then told to resume 'own visual navigation'. I politely informed ATC that I was now uncertain of my position.

Taking off from Binbrook, England on March 2 1990. Very cold weather, and icy runway.

After three transponder changes and a fair amount of ATC assistance I proceeded towards Ely. Similar instructions and transponder changes occurred all the way down through England, and as I passed the Honington Military Control ATC called up: 'Hotel Papa Hotel. You have two Tornados at 9 o'clock.' I turned to look out over the port wing and saw two swiftly moving fighters flying in tight formation.

I replied, 'Hotel Papa Hotel, I have traffic sighted.'

Back came AT's reply: 'Good, sir. You are now in Indian Country.'

I chuckled. The helpful, good-natured attitude of the ATC throughout the whole leg had been of great comfort.

I flew across the Thames and over South Bend. I was starting to feel comfortable. I was clearly on track, I could see the English Channel, the sun was out, and the feeling was coming back into my hands and feet. I glanced down. The countryside looked beautiful. There was no snow down here, but the wind still appeared to be strong. It was all working. I was in control.

Then the crankshaft snapped ...

II

The Police Superintendent's office was small and warm; I was numb. The Superintendent on the other side of the desk was patiently quiet as I sat slumped forward in my chair, head in my hands.

What was I to do? All seemed finished. I'm not sure how long I sat there—I suppose it wasn't all that long, but after a time my thoughts were transferred to an incident that had occurred some 12 months previously. A close friend tragically lost his son in a motor vehicle accident. As we embraced in grief, he placed one hand on my shoulder, lifted his bowed head, looked at me and said, 'Eddy, no matter what happens to you in this life, nothing is worse than losing your son.' The pain and suffering in his eyes are things I will never forget.

The memory threw me back to reality. Mine was a minor accident by comparison! I looked up and sat back in my chair. I think the Superintendent was relieved! At that precise moment the telephone on his desk rang.

'Superintendent,' he said. 'Why, yes, he's sitting right in front of me now. It's for you, sir.' He passed the old black handpiece across to me.

'Hello?' I mumbled. I was finding it difficult to speak.

'Brian, it's Harry Shaw here. What rotten luck, just heard about it. I've arranged for a team to pick up the plane and get it to Headcorn. It's a little field about five miles away.'

I was now thinking as clearly as I should be. I stood up and thanked the Superintendent, and asked if it would be possible to get a lift back to the field. There was work to do.

On arriving back at the field, I was greeted by a team of about 15 young fellows from the Headcorn airfield and nearby village. They had brought a truck with them, a little small, I thought, for holding Matilda, but it was a start. We walked across the paddock to where my poor yellow biplane lay at an odd angle, with the undercarriage embedded in mud.

We slowly pushed her across the field to the gate and then discussed ways and means of lifting her onto the tray of the truck. One of my helpers said he had just the piece of equipment we needed, and trotted off down the lane. About 20 minutes later he returned, driving the most extraordinary vehicle I have ever seen. It was an old World War I armoured car, which had been modified somewhat for farm work. The large caterpillar tracks handled the mud in the field well and the crane-like structure that protruded from the front of the cabin looked strong enough to lift Matilda. Some two hours later we had Matilda precariously perched on the back of the truck, her wings drooping at an untidy angle over the edge of the tray.

We all sweated and cursed for eight hours negotiating the narrow lanes, but eventually the airfield gate was in sight: only 500 metres to go. Unfortunately, however, this last stretch had four large, overlapping oak trees.

There was no chance of getting through. 'We'll have to go around the back way,' yelled the truck driver.

'How far?' I asked.

'Three miles.'

'That's too far—it will take another five hours. The wings will have to be folded.'

'That's not possible,' said another of my helpers. 'The Bureau of Air Safety have to see the plane as it is.'

My reply was short and to the point. The wings were folded back, the upper and lower wings being propped apart by a couple of branches from a nearby tree, and we continued for the last short stretch.

I had come down at midday, local time. By one the following morning we had Matilda in a hangar at Headcorn.

Jenny and Tony, when I managed to contact them, had heard about the mishap over the radio. They were making their way back from Binbrook, Jenny to her sister's place in East Grinstead and Tony to his mother's just out of London. My confused and somewhat garbled messages had reached them.

At around 2 am, Jenny, Glen and Mike came through the door into the small briefing office, where I had been sitting for an hour or more desperately going over the disastrous day's events. There was little to say. We all made a very dismal and melancholy sight. Mike dropped his large frame into the chair next to me.

'Brian, Scottish people usually find a cheer in a little drop of good whisky, no matter what the circumstances,' and with that he took out a small bottle from his pocket—his best Glen Fiddich.

Tony arrived, and we walked briefly around Matilda. She looked terrible—the folded wings didn't help. There was obviously nothing more to be done that night, and it was time to get some sleep. It was a quiet drive to Mike and Glen's place—what was there to say? A major disaster on the first leg. I climbed exhausted into bed at 3.30 am.

Early the following morning Tony and I were back at the airfield. On inspection, we found the left landing strut was damaged, the left lower undercarriage bent, and ribs broken in the tailplane. The nose cowling was torn and dented where the propeller had come backwards across it and had then smashed through the upper wing, splitting the main spar. Pieces of the propeller had been found close to the field where I had come down. The engine, as expected, was finished.

I slowly walked around Matilda. The mud-splattered fuselage and wings all added to what made a very dismal picture indeed. It was hard to try to be positive, but I just had to be.

I looked up to see a small, slightly-built lad walking through the open hangar door across to where Tony and I were standing. Stretching out a hand in greeting, he said, 'Hi, I'm Jamie Freeman. My father, Chris, owns the airfield. Bad luck about yesterday. Would you like to come for a fly?'

We flew in his Piper and retraced my flight track from South Bend to the point of engine failure. Five minutes earlier I would have been over South Bend, five minutes later over the English Channel.

Later in the morning two members of the de Havilland Moth Club landed at the airfield in their Tiger. Mike Vaisey and Jim Norris were to play a major part in getting Matilda back into the air in three weeks.

That afternoon I flew with Mike in his Tiger up to Hatfield to inspect hangar facilities, and was given permission by British Aerospace to hangar Matilda there while repairs took place.

The next three weeks were a strange period for me. The initial problems of arranging repairs, obtaining a replacement engine and getting insurance assessments were resolved relatively quickly. Stuart McKay, General Secretary of the de Havilland Moth Club, fortunately had a spare engine that I was able to borrow, and the Australian Aviation Underwriting Pool, with whom I was insured, very quickly assessed the damage and enabled the repairs to get underway immediately. I was very grateful for this.

I received many phone calls and letters from home during this period, all offering support and encouragement. I remember in particular a call from Sir Fred Chaney, our patron: 'Fred here, Brian. Bad luck about the crankshaft; just put it out of your mind and get on with it.'

Simple, to the point, and, of course, correct. All the calls and letters helped a lot, as they reminded me I had to remain positive. I was also fortunate to have Jenny there and to be at Mike and Glen's place.

There was no doubt that during those three weeks I was fairly difficult to live with. One particular morning I remember well. I was about to leave to drive up to Hatfield to check on how the repairs were coming along, when Glen walked into the sitting room and handed me a small brown envelope. It was addressed to me, with the instruction 'Please forward.' Inside was a small card, which said:

To you, Brian Edwards, who will be remembered in Holy Mass offered on your behalf for the Intercession of the Holy Martyrs of England and Wales.

I was astonished. Not being of the faith, I couldn't imagine why anyone would do this. We tried to find out where the card had come from, and attended a Catholic Mass at a small church in East Grinstead that Sunday, but we were unsuccessful at finding anything out. Both Jenny and I were moved by the act.

As far as the repairs were concerned, I knew everything was going as quickly as possible. Tony, Mike and Jim worked virtually around the clock, but test-flying had to be done again. The new engine and propeller had to be thoroughly tested, as did the fuel transfer. Although there appeared to be no damage to the fuel system, I had, after all, come down a little more firmly than usual.

There was also the possibility of my visa and clearances for some countries expiring. If this happened, it would take months to renew them. Also, the weather patterns in the Middle East, India and South-East Asia would start to change for the worse. The monsoon period would start developing towards the end of May, and this would not be all that satisfactory for Tiger Moth flying.

During this time my frequent trips to Hatfield to check on progress of repairs were mixed with a feeling of urgency about getting going again.

I endeavoured to keep fit by jogging every morning. There was little more I could do concerning the route preparations. Everything had been done thoroughly before. There was still one clearance to come—through Pakistan—but this was not considered to be a problem, and would be available by the time I was to leave Muscat. I went up to London to do a little teaching at the Royal Free Hospital in Hampstead, but I could not relax. I had prepared for what I thought was every possible contingency, but not for a broken crankshaft.

My constant thought was that all of the problems— weather, navigation and mechanical—were still to be encountered. Sleepless nights were not uncommon. Jenny was a great comfort in those difficult times.

The dull, drizzly days seemed to add to the intermittent depression that would occasionally engulf me. So many people let down—I just had to get on my way.

Wednesday, March 21, Matilda was ready to test-fly. Tony, Mike and Jim stood by anxiously as I strapped myself into the cockpit. Tony pulled the propeller, and the engine started instantly and sounded good. I taxied out at the grass apron, and called Hatfield tower for permission to take off. I flew cautiously in the circuit for a while and then ventured out over the countryside. Everything seemed fine. Tony, Mike and Jim had done a great job and there appeared to be no problem with the fuel-transfer system.

The following day, March 22, I was out at Hatfield for what I hoped was the last test flight. The team had worked almost all night to readjust the tappets and check the carburettor settings. I flew Matilda again, and everything seemed fine.

I taxied back to the hangar late in the afternoon, having completed a fairly strenuous six hours' test-flying. It looked as though I would be held up waiting for my new propeller as the one I had been using for test-flying was really not suitable for such a long flight. The weather at that time was reasonable, and the following day's forecast was for much the same. However, a large low was sweeping up from the south-west, and the forecast for the weekend was bad—rain and gale-force winds. If I didn't get away the following day I would probably be held up for another week.

Hatfield, March 22 1990—the day before take-off.
From left: Brian Edwards, Mike Vaisey, Tony Wilkins, Bill Norris.

I leaned forward and turned the switches off. The engine grumbled to a stop and I sat back in my straps, rested my head in my hands and closed my eyes. Damn the propeller!

I was startled by the sound of someone shouting. I turned around, and there was a smiling Tony, holding my brand-new propeller.

'It's arrived!' he yelled.

I didn't bother to get out of the plane. The team swapped the propellers over, we started her up again and I taxied straight back out and did another two hours' flying. Matilda performed beautifully, and when I finally taxied back to the hangar it was dark.

But I was elated. I was ready for departure. We had made it in three weeks. Next day, March 23, I was to leave on my first leg for the second time.

Jenny and I stayed the last night in a small hotel close to Hatfield. I was at the airfield at six the following morning. The weather looked reasonable, and was forecast to improve as the day went on, particularly over France. My track was to take me over Lydd, Cape Gris Nez, Le Touquet, Mantes, and across the Seine to Villacoublay, a French airforce base just out of Paris, my first overnight stop.

I submitted my flight plan, and walked down from the tower to my aircraft and the small group of friends who had come to wave me off. Mike, Glen and Blair had joined us for breakfast that morning and had come over to the airfield to farewell me. Stuart McKay from the de Havilland Moth Club, Ian Wooldridge from the Daily Mail, a great supporter, and a few people from British Aerospace gathered around Matilda.

I was very nervous, but tried to hide it by busying myself with my life jacket. It was a tight fit over my sheepskin flying jacket, though, and I made a bit of a mess of it and in the process managed to drop and break my glasses. Glen, seeing what had happened, ran forward and pushed hers into my hand.

'Will these do?'

'Oh, God, I suppose so.' I thrust them into the top of my flying jacket. What a time to break my glasses! Glen's were fine, however, and served me well all the way to Australia. Farewells were said. Tony was to travel by commercial jet and meet me in Athens, Calcutta and Singapore.

I strapped myself in the cockpit, helmet and goggles on, map in hand. Jenny came up and gave me a 'goodbye' kiss: 'All the best, darling, see you in Perth.' She was battling to hold back the tears, but was putting on a brave face.

Tony swung the propeller. Matilda started first swing. The Gypsy Major sounded beautiful despite her 40 years. As I was about to taxi I felt a tap on the shoulder. It was Mike Vaisey, who had come up to the cockpit He shouted something, but the slipstream from the spinning propeller and the noise of the engine drowned out his voice. I turned around with some difficulty, as the harness, sheepskin jacket, life jacket and cramped cockpit left little room to move. He cupped his hand over his mouth and leaned into the cockpit.

'I said, this could not have happened if it wasn't for your father.'

What could I say? I felt a distinct lump in my throat. It was true, but I'd never thought of it like that before, and I felt strangely uplifted. With moist eyes I called the tower and obtained a clearance to taxi, pulled my goggles down, bumped across the grass to the runway and lined up.

'Hotel Papa Hotel ready,' was my call to the tower.

'Hotel Papa Hotel—Hatfield tower, clear for takeoff, good luck, sir.'

Smoothly opening the throttle, I gently pushed the stick forward, held the right aileron into the wind, lifted off and climbed into the dull overcast sky. I turned left at 500 feet, took a heading for Lydd and continued to my assigned altitude of 2000 feet.

I was on my way to Australia!

Chapter 4

From Hatfield to Rome

I cruised at 2000 feet down to Lydd, then overflew and took up a track across to Cape Gris Nez on the French coast. The cloud was low, but visibility was fairly good. It was only 22 miles across the Channel and simple navigating, but I was relieved to reach the coast of France. I transferred my radio frequency from Lydd to Le Touquet. Although English is the international language used by all air traffic controllers, it took a little time to get to understand English spoken with a French accent. It was just as hard for the French to understand my English with an Australian accent. We managed, though, and they were most helpful.

I went down the coast of France and over the mouth of the Somme to my last coastal fix, EuMers. Then inland to Mantes, over the River Seine and finally to Villacoublay, a French airforce base just out of Paris. The French countryside was quite beautiful and green, and all in all it turned out to be a simple navigational exercise. Nevertheless I was relieved to have arrived safely and without incident. The disastrous start three weeks before was still very much in my thoughts.

Arrival at Villacoublay, March 1990.

On my leg from the coast across to Villacoublay I had noticed a little extra play in the control column, and while refuelling I mentioned it to the Commander, who provided a mechanic to check it out. We both spent an hour leaning into the cockpit, and, although we could not entirely resolve the problem, I was satisfied there was no risk involved in continuing. Having checked and tied Matilda down, I was then escorted to the Officers' Quarters.

On settling in, I was surprised to find that there were no other personnel there. Most of the base had gone home. I was shown to my room, which was very comfortable, dropped my kit on the floor and fell onto the bed. I had left Hatfield at 9 am, and it was now 5 pm.

I rang Jenny, who was excited to hear from me. But I was somewhat subdued. I had successfully completed my first leg, but I still had a long way to go. It had to be one day at a time, and I could think no further than the next day's flying.

After the call I took out my maps for the following day and went over the landings, frequencies and fixes. It would be a relatively simple leg down through France to Toulouse/Francazal, another French Air Force base. Everything seemed to be in order.

I then went down to eat. I was ushered into an empty dining room. All the chairs and tables were upturned, with the exception of one table in the centre of the rather large room, which had one place set, one chair and a vase with one rose in it. Being the only one in the dining room, I had my own personal waitress, which made me feel a little embarrassed. She was a rather small, dark-haired girl, dressed in a semi-military uniform. She had that attractive untidiness about her appearance that only the French seem to do well. She couldn't speak English and I couldn't speak French, but her flashing smile put me at ease. The menu was simple, and the meal was accompanied by a small carafe of beautiful burgundy. After a leisurely dinner I returned to my room.

Flight planning had been done soon after landing, ready for an early-morning take-off. All that was necessary now was to get a weather report first thing in the morning, adjust headings and time intervals, fill out my small personal flight plan, and clip it to my knee pad which contained the appropriate aerodrome diagrams—then I'd be ready to go. I slept soundly.

At 4.30 am I was awake. A quick breakfast, and then down to the hangar. The weather report was favourable: clear sky with no cloud—CAVOK in flying jargon—wind a little strong, 25 to 35 knots for parts of the trip, but it would be on the tail.

I felt nervous walking out to Matilda. Here I was on Day Two, with still a long way to go. But she looked fine. I packed my kit into the small locker behind the cockpit, untied her, checked the cockpit, and left her chocked, fuel on, throttle closed, switches off. I walked around to the front of the plane and pulled the propeller through. Although there was quite a crowd to see me off, no one was familiar with hand-starting a Tiger Moth. Throttle set, switches on, one flick of the propeller, and she started first time.

Climbing into the cockpit and strapping myself in, I realised how much these procedures were to become part of my life for the next six weeks.

Just as I was about to taxi out, having obtained a clearance, I glanced over to the crowd and waved. Then a small, duffle-coated person pushed through the crowd and ran to the side of the cockpit. It was my waitress from the night before.

'Bon voyage, Monsieur Edwards,' she shouted, and thrust a bottle of wine into the cockpit. It was a very nice gesture, but I had nowhere to put it! It meant, of course, stopping the engine, climbing out, putting it into the back locker and going through the whole procedure again—all up, 30 minutes! Nevertheless, I appreciated the thought—and the wine!

Then I was off. Although it was quite cold, zero degrees, the sky was beautifully clear and crisp. Some fog had been forecast en route, but it was to clear by 9 am local time and I doubted it would present any problems. My track to Toulouse/Francazal was almost due south, and I didn't expect the visual navigation to be all that difficult.

I should have known better. The many small, snow-covered towns were very difficult to recognise and the roads were impossible. I struggled to find Limoges, my first major fix, but eventually the airfield came into view and I landed to recheck headings, and, I must confess, to regain my confidence.

Thirty minutes later I took off again, and continued to Toulouse. The visual navigation continued to be difficult, and, on picking up a small airfield, Soulat, I decided to land again. The chief flying instructor of the small flying club that operated out of there discussed the route to Toulouse with me. We had a cup of tea together, he swung the propeller for me, and I departed for Toulouse, landing one hour later—to be greeted by the Base Commandant, Colonel Bigot, his wife, and the local TV crew. The Colonel, with the expected military but polite demeanour, and his wife, with her gracious manner, made me very welcome. They later wrote an article about my trip for the local paper.

After the television interview I was shown to my very comfortable quarters in the officers' mess, and made a quick call to Jenny to say everything was OK. I sat on my bed, feeling quite dejected about the day's

Toulouse/Francazal, France.
From left to right: Brian Edwards, Colonel Bigot,
French Air Force helicopter pilot.

flying. The weather had been good, but my navigating hadn't. Although my total flying time had only been four and a half hours, I had made two unscheduled landings to be sure of my position. This in itself was not a bad thing, as on both occasions I was not really off track, and it was good airmanship under the circumstances. But on one occasion I had lost my confidence a little, not seriously, but enough to develop a mild case of moist palms and a dry mouth.

I had divided each leg into 20-mile sections, and, depending on the wind strength and direction, this usually meant about 20-minute time intervals, yet I had failed to get position fixes. I had made what was to be a simple navigating leg in good weather into something of a trial. There were far more difficult legs to come—I could not let this happen again. In a somewhat melancholy mood I took out my maps to plan for the following day.

The next morning I was greeted on waking by a dull, overcast sky. The cloud base was about 1000 feet and a mist-like rain was falling. It was not going to be a good Tiger Moth day. I was met by an officer and taken to the mess for breakfast. He spoke English quite well, and was an experienced pilot. Over a breakfast of coffee and croissants with his family, whom he had brought along to meet me, we discussed my leg to Cannes. He explained that he thought the weather would lift by lunchtime, but that the wind might be a problem once I reached the coast. A 35-to-40-knot wind from the south west had been forecast—the Mistral.

The weather cleared by lunchtime, and I was farewelled by the colonel, his wife and my officer friend. After making very good time across the coast with the wind on my tail, I turned east and tracked via the coast toward Marseilles.

The Mistral did not disappoint. The wind blew violently, forcing me out to sea all the time. I fought desperately to keep on track, continually changing headings to maintain coastal navigation. The wind speed must have been in the order of 40 to 50 knots, but once I was past Marseilles it lessened and I reached Cannes comfortably before nightfall. I was able to find a corner in a hangar for Matilda, and I booked into a small hotel near the airport.

That night I met two other pilots who were flying a light aircraft to Australia, an American and a New Zealander. The New Zealander, an experienced pilot, was a large, untidy man. His gruff and abrupt manner concealed a rather gentle and helpful character. Over the next 48 hours we had a number of useful discussions regarding the navigational problems involved in flying to Australia.

The American, on the other hand, appeared to be more concerned with 'sight-seeing' than flying. They had departed the UK at about the same time as I had, but were flying a more modern aircraft equipped for instrument flying. It was good to talk to them, but I must admit one incident surprised me.

I was in my room preparing my maps for the following day when the American pilot came to visit me. I had the maps spread out on the bed, and was checking the routes with my protractor, a small rectangular instrument that gives compass bearings for particular tracks. He peered over my shoulder and began to ask questions about the route I was to take. On many legs it was to be similar to his. He asked me what I was doing, and I explained to him briefly that although I had previously prepared my tracks and headings, I always rechecked each leg every night. I was holding my protractor at the time, and he offhandedly asked what it was. I thought he was joking and glanced up, expecting to see him laughing. But he was not: he was serious! When I replied that it gave me compass bearings of my track, his only comment was, 'Gee, that's neat!'

Unfortunately, as I found out later, the pair did not complete their trip, and abandoned it and their plane in Chittagong.

The weather was bad again the next day: heavy rain and thunderstorms en route to Rome. I spent the day in my hotel room going over maps.

The following morning the weather had improved, and I was cleared to proceed to Rome Ciampino. My route took me mainly along the coast, so I expected the navigating to be straightforward; the only problem, as usual, would be the weather. A small front had been forecast, and I would probably hit this at Savona. All in all, though, I was quite looking forward to flying this leg down the Cote d'Azur and past Monaco—places frequented by the rich and famous. My initial clearance, being low at 1000 feet, would make the first part of the trip quite interesting.

I took off and climbed out into a dull, cloud-drenched sky towards Cape d'Antibes and Cape Ferrat. Ahead of me there was a thin bank of cloud at about 1000 feet. I noticed that the sea below me appeared to be a little rough, and as I turned over Cape Ferrat the horizon ahead looked quite dark. The wind speed continued to pick up and the turbulence also increased.

It began to rain. Visibility kept on falling, and I was down to 500 feet. I didn't want to lose sight of the coast, but the turbulence was much worse close to the coast: the down draughts off the mountains were severe, making it very difficult for me to maintain control. Because of this I had to fly further out to sea than I would have liked. The rain was heavy, and water was streaming off the leading edge of the wings and the tip of the propeller, giving me the impression that I was flying in a tube with glistening walls.

It rained continuously for the next three hours, but the turbulence of the mountains decreased as I passed Savona and the wind came around onto the tail. The heaviness and persistence of the rain, the low cloud base and the poor visibility kept me flying very low, but I could quite easily see the coast and felt confident about my navigation.

However, the rain was coming into the cockpit, trickling across the instrument panel and splashing over my goggles despite my large windscreen. As well as this, it was seeping through the cowling, and running back along the cockpit floor, which had become awash.

I was very wet, but of course the main problem was that the rain was soaking my map, which I was frantically clutching in my left hand. Slowly but surely it was becoming a pulpy mess, and therefore difficult to read. I cursed that I had decided not to place it in the plastic cover beforehand. My radio was also wet, and transmission had become impossible.

I loosened my harness a little, which allowed me to bend forward in my seat and hold the map further up into the front section of the cockpit. This kept most of the rain off, and although I was in an uncomfortable position my map remained in one piece.

The rain eased as I got closer to Rome, but it looked as though thunderstorms were building up to the east of the Eternal City.

Thirty miles out, I called Rome Fiumicino control, and fortunately my radio transmission worked. I was cleared across to Ciampino without any delay. The thunderstorm continued to build up, and looked very ominous, but I located the airport fairly easily and was given clearance to land. The wind was gusting to 15 knots and as usual was directly across the strip, but the coming thunderstorm meant it was time to get on the ground quickly.

When I landed, my slightly dry mouth was in stark contrast to my soaked flying suit and sheepskin jacket. I was given parking instructions and directed to the other side of the airport, some 1000 metres from the hangers.

I had expected to be met here, but no one was around. I climbed out of the cockpit, thankful to be on the ground, and looking forward to a hot shower and dry clothes. I glanced across to the mountains and saw the thunderstorm approaching. The wind was freshening. Matilda wasn't going to survive this.

A routine call to the tower requesting hangar cover was met with a mixture of humour and disbelief. I quickly explained that my fabric-covered plane would be seriously damaged by the hail. They understood this. I swung the propeller and taxied a little faster than I should have around to the other side of the strip. On reaching the first hangar, I was greeted by two smiling faces: Sandro Rosati and Mauro Amaded, from SNAM, an Italian air charter company. Where had I been? They had been waiting all day; they hadn't seen me land.

They hurriedly pushed Matilda into the hangar just as the thunderstorm hit. Lightning struck the tarmac close to where I had previously been parked, and hail fell. Had I been out in it, Matilda's fabric would have been seriously damaged, probably beyond repair.

The weather really closed in for the next two days. One front followed another.

My friends Sandro and Mauro were marvellous hosts, but I found it very difficult to relax or socialise. They had booked me into a lovely hotel up in the hills, Hotel Cabalo. From my room I could see down over the valley; it was a beautiful setting, but my thoughts were elsewhere. The flying ahead of me, the navigation through the mountains and my impending first long-distance water crossing, over the Adriatic Sea, constantly pushed their way into my thoughts. And the weather made it worse. I spent time tinkering with Matilda.

I also spent some time with SNAM's chief pilot, Moreno Caso. We talked at length about the weather patterns in the Middle East and the weather I was likely to encounter. Favourable winds were probable, but sandstorms might be a problem. I was very grateful to be able to talk to him; his advice was invaluable.

Jenny, who had now returned to Australia, phoned and told me that everything was fine at home, and Michael, Christopher and Susie were closely following my progress. It was a great source of inner strength for me to be able to talk to them—I doubt they will ever realise how great.

It was in Rome that I began to get an awkward feeling of uncertainty and doubt. I didn't quite understand what it was—I still don't—but it was to surface a number of times over the next six weeks. It only really came to me when I was on the ground. The feeling had first arisen during the three weeks in England following the engine failure: an odd mixture of uncertainty, doubt, probably fear—I don't really know. One thing I did know was that sitting on the ground was no good for me; I just had to get on my way.

After two days there was a break in the weather—just enough time for me to get out down the valley and across the mountains to Brindisi. Roberto accompanied me for the first 30 miles in his own plane, a Rondone II, a lovely little aircraft. I was quite pleased to have him fly with me for a short period, and we flew in loose formation down the valley. After about 30 minutes he drew up ahead of me, waved and peeled off to return to Ciampino. The company boosted my confidence a little.

The awful feeling of uncertainty slowly eased away as I got on with the job in front of me. The sky was clear, and the navigation was turning out to be not all that difficult.

I flew down through the valleys, flying below the cloud and trying to pick up the motorways. It really was quite pleasant and only a short leg, some 260 miles, to Brindisi, where I planned to refuel quickly and get on my way to Athens.

Chapter 5

From Brindisi to Luxor

I arrived over Brindisi Airport, having made good time. I landed and refuelled, then proceeded to flight-plan for my Athens leg. Within the space of an hour, however, the front that had been chasing me from Rome caught up. Rain fell and the wind blew. The airport was closed. By the time the front had passed over it was too late to leave. I was furious. Another day gone, and only 260 miles covered.

I found a small hotel close to the airport, climbed the usual two flights of stairs and was ushered into a dingy room. Paint was peeling off the walls, but a small bed and a curtain-covered wardrobe made it adequate for my purposes. I began to prepare my maps for the following day. Brindisi to Athens—it didn't seem a difficult leg, but it contained my first major water crossing. It would probably be quite an eventful day.

It was difficult to get to sleep, and I tossed and turned for some hours. Finally I got up, pulled on my jeans and boots and tramped down the two flights of stairs and into the small, narrow street in front of the hotel. The air was cool, in stark contrast to my stuffy, hot room. I walked the narrow streets of Brindisi for about two hours. There were still quite a lot of people in the square, even in the early morning. The walk did me a lot of good, and on climbing back into bed I fell asleep and slept soundly for the next couple of hours.

I awoke at 4.30 am, and after a quick breakfast packed my maps and was at the airport for a 6 am departure. The weather looked good—a cloudless sky, and a gentle five-knot breeze blowing off the ocean.

The weather report indicated fairly uncomplicated weather between Brindisi and Athens. Matilda started easily with one flick of the propeller, and I launched myself out over the Adriatic on course for Corfu, some 130 miles across the water.

A forecast 20-knot southerly wind meant that to maintain my track I would need to set a course somewhat into wind; that is, I would have to head further south to maintain my intended track. I calculated a heading that would give me my required track, and climbed to my assigned altitude of 2500 feet. The sky was clear, and unfortunately I allowed myself the luxury of feeling comfortable.

I concentrated intensely on keeping an accurate heading, knowing that without positive visual fixes assessment of drift would be difficult. I glanced down at the brilliant blue ocean, but there was something different about it. It appeared that the wind was coming from the north, not the south.

Brindisi control couldn't be raised, so I decided to descend to assess the wind accurately. I dropped down to 100 feet and confirmed that the wind was coming from the north, at about 15 to 20 knots.

It hadn't occurred to me that I might have to do this, but I suppose that for someone who has done a bit of surfing and rowed surf boats assessing the wind over the water is quite easy. Surfing had been part of my life for many years. Like most Australian children, I had learnt to swim at an early age, and spent a great deal of time on our beautiful beaches. The ocean was an integral part of my life.

My satisfaction at being able to assess the wind was coupled with the dreadful realisation that I had been on the wrong heading for about an hour and was now well south of track. I climbed back up to 2500 feet and tried Brindisi—no go. I set about changing course, estimating that I was now about 20 miles to the south of my track. I had one more go at Brindisi, but to no avail.

Then a very English voice crackled over my radio: 'Aircraft calling Brindisi—this is Speed Bird 401. Say again call sign and I will transmit your message.' British Speed Bird 401 was a British Airways 747 cruising at 35 000 feet above me.

'Speed Bird 401, this is Hotel Papa Hotel, request revised wind direction and speed below 5000 feet between Brindisi and Corfu.'

Speed Bird 401 called back in a few minutes. 'Hotel Papa Hotel, Speed Bird 401: unable to contact Brindisi. Can we help?'

'Speed Bird 401, Hotel Papa Hotel: thank you. Do you have wind speed and direction below 5000?'

'Well, we can certainly give it to you at 35 000 feet.'

'Not much help to me.'

'No, I suppose not. What height are you flying?'

'Two thousand five hundred feet.'

'Why?'

'That's my assigned altitude.'

A few minutes' silence ensued.

'Hotel Papa Hotel, Speed Bird 401: say again altitude and what is your true air speed.'

'Two thousand five hundred feet and 70 knots,' I replied.

A few seconds' silence followed.

'Hotel Papa Hotel—Speed Bird 401—say again true air speed, true air speed.'

'Speed Bird 401—Hotel Papa Hotel—70 knots; seven zero knots, 70 knots.'

'What type of aircraft are you flying?'

'Tiger Moth.'

'Good God! What's your destination?'

I didn't have the heart to say Australia, so I said, 'Athens.'

'You must be mad,' came the reply.

I decided that didn't need an answer.

He gave me a wind direction that agreed with my own assessment. I thanked him, and he wished me all the best. Speed Bird 401 then moved out of range.

I flew on with my adjusted heading, desperately hoping I was on track, and searched the horizon for some sight of land. After about two hours a thin band of grey-green that looked like land came up on the horizon. It was land. I hoped it was Corfu. It was. My heading adjustment had been correct; I was elated. My first long water crossing had been completed successfully.

I decided that I just had to sing to celebrate, and have a sip of water. I sang the first verse of Waltzing Matilda and then settled back down to continue my navigation around the coast of Greece to Athens.

The scenery was quite beautiful; the mountain greenery came right down to the sea, leaving little room for beaches, and visibility was good. The wind picked up as I got closer to Athens, particularly in the Gulf of Corinth, and by the time I was over Athens there was a strong crosswind blowing, gusting occasionally to 25 knots. I was cleared to land and slotted in behind a 747—not Speed Bird 401, however!

As I came in on short final, I began to have doubts about being able to put Matilda down properly. Full opposite rudder and aileron into wind still did not keep me in the centre of the runway. I executed a go around and tried again—still no good. Then I was told to hold out over the Bay, which I did for one hour. The wind did not lessen, and I explained to ATC that I thought I could land on one of the taxiways into wind. I would need only 50 metres. Could I have a clearance? All hell broke out in the control tower. I'm sure they thought I was a 747!

After considerable explanation I was given a clearance. In fact, I only needed 25 metres! I had been seven hours in the air and I was pleased to be on the ground—pleased but tired.

Tony met me here, and it was good to see him. I felt Matilda was performing well: fuel and oil consumption were good, and handling characteristics appeared to be fine. I decided to have a day's rest in Athens while Tony checked Matilda out, and I spent most of the day asleep in my room. About 4 pm Tony returned to the hotel and woke me with a cup of coffee. We sat drinking the scalding coffee and discussing the next legs, particularly the ones through Egypt and the Middle East.

'You know, Tony,' I said offhandedly, 'I should have organised a proper cover for the cockpit before I left England. I doubt I will have hangar cover from now on.'

With that I stretched back on the bed. I was more tired than I thought. I must have dozed off, as I was again woken by Tony. He was covered in dust and cement, and was carrying a large piece of blue plastic.

'I picked this up from the building site across the road—I don't think they'd miss it. It should make a good cover for the cockpit.' It did, and was with me all the way to Australia.

I took off on April 2 for Iráklion, Crete. Strong winds were forecast, but generally the weather was reported to be good, and I planned to fly the 154 nautical miles to Iráklion by following the small islands that spread themselves all the way there from Athens.

I departed Athens at 7 am local time, but had to return after only 30 minutes because of radio problems. I was on my way a second time at 12 noon, and arrived in Iráklion three hours later. The islands had made the navigation very easy.

Here I met a group who were flying from the UK to Australia in some 25 aircraft, all of them 1950s construction. It was good to chat to them about their experiences. There did appear to be some internal problems regarding the clearances and route, and I must admit I was rather pleased to be flying on my own, without having the problems associated with trying to get 25 pilots to agree about such things as route segments, clearances and the like.

The following day I departed for Alexandria—270 nautical miles across the Mediterranean. The weather looked good and I was now a little less apprehensive about flying over water. I tracked around the eastern end of Crete and then took up my heading across to Alexandria. I had decided to make a 'positive error' so as to hit the coast of Egypt west of Alexandria. This would mean I would then only have to turn one way east from El Dabba, some 70 nautical miles due west of Alexandria.

I cruised at 2000 feet for most of the way, dropping down to 200 every so often to check on wind speed and direction. I was surprised that the leg, over a brilliant blue sea, was so trouble-free and enjoyable.

Occasionally I would let my mind drift back to surfing at Gracetown, a small coastal settlement 250km south of Perth where my family and I had spent many wonderful holidays. The water was the same blue, but the waves were bigger at Gracetown, especially at North Point on a good day.

I arrived without much difficulty at El Dabba, and then tracked east to Alexandria. On arriving in the circuit area, I found that, as usual, the wind sock was horizontal, showing a wind direction directly across the strip, but this, I felt, was not a problem anymore. As far as Tiger Moth flying was concerned, I could now reasonably class myself as being fairly hot!

I landed, however, in a rather untidy manner, managing to bounce twice before finally keeping her on the ground—always an embarrassing thing to do. I coped with the usual trials and tribulations of customs and refuelling. Although the whole leg had taken only five hours, the four hours it took to refuel and clear customs drained all my energy.

When I finally arrived at my hotel I was exhausted. I had just enough energy to check my route for the following day and drink a couple of bowls of soup before falling into bed in my flying suit, too tired to change.

I had been a little nervous on leaving Iráklion. The 270-mile flight over the water was a long way without a visual fix, and the possibility of getting off course was very real. Ground speed and drift are easy to calculate over land, with visual fixes coming up every 10 or 15 minutes. This, of course, is not possible over water, and I suppose the concentration required for trying to assess wind speed and direction took its toll.

I was awoken with my usual 4.30 am wake-up call, had a hurried breakfast of coffee and toast, and reached the airport just as it was getting light. I carried my kit over to the control tower and began to climb the now familiar two flights of steps that rose in front of me. Cursing a little, as my kit seemed heavier this morning, I struggled up the steps.

Suddenly there was an extraordinary shout, a mixture of fright, pain and annoyance. I had stepped on a sleeping Egyptian! It must have hurt, as my boots were steel-capped, and to make matters worse I had fallen forward, dropping my heavy kit on him as well. We swore at one another. (Although I couldn't understand him I'm sure he was swearing!) Then the uncovered bundle sat up, moved down one step, rolled over, and, I assume, went back to sleep.

I managed to find my kit, and continued my climb, this time navigating my way around the sleeping bodies that I could now see.

On reaching the control room, I was astonished to find what appeared to be a whole family asleep inside. Dad very sleepily accepted my flight plan, wished me luck, and then lay down and went back to sleep.

Despite the unusual goings-on in the tower, I was cleared over to Cairo quite quickly and granted permission to fly direct to Luxor down the Nile valley at 1500 feet. This certainly was a special clearance, as flights are usually not permitted in that area—particularly at that height.

It was an unforgettable experience. The Nile valley was brilliantly green, flanked on either side by the barren yellow desert. An occasional boat with the unusual triangular sail drifted unhurriedly along. As I got closer to Luxor the magnificent pyramids came up on my starboard side. This was a truly memorable sight.

On landing I was met by my handling agent, a small, narrow-eyed man with a flat nose and a large black moustache. Despite his unappealing appearance, he was a pleasant fellow, and very helpful. He repeated his name a few times, but I couldn't understand him. 'Jim?' I suggested. He nodded, smiling. So 'Jim' he became.

Once settled in my hotel, I began the usual ritual of going over the route for the following day. It was to be fairly long—475 nautical miles—and the navigation would be difficult. I had 200 miles across the desert and then 275 miles across the Red Sea to Jedda.

The main problem would be leaving the Egyptian coast at an accurate fix, and then assessing the wind over the ocean and being able to pick up a fix on the Saudi coast. I realised I would have to use the positive error approach, and calculated a heading that would give me a track allowing me to hit the coast with only one possible turn to reach my destination. I sat pondering this for a couple of hours and decided that a small headland, Ras Bonas, would be my final fix on the Egyptian coast. I would then track out across the Red Sea, laying off sufficiently to allow me to hit the Saudi coast 50 nautical miles to the north of Jedda. On the chart there appeared to be a small river system at this point, which I felt I should be able to pick up. But it was not going to be easy.

The awkward feeling of uncertainty again engulfed me. It hadn't happened for a while. Would my navigation be accurate? Would I be accurate with my assessment of the wind? What if the weather changed, particularly over the Red Sea? I was getting more and more uncomfortable.

I rang home, and managed to get through. It was comforting to speak to my family. They were all excited about my being in Luxor: 'What an exciting place!' I really felt as though I might have been sitting in a roadhouse on the Nullarbor Plain! I remember getting a call from a friend, Peter Burns, that night. Had I visited Karnak? I think he was disappointed when I replied that I hadn't, and didn't really plan to. I appreciated the call, though.

I couldn't relax, and was still awake at 2 am. I was getting into a state. The navigation would be difficult—how would I manage to cross the Red Sea if the weather turned foul? I got up and sat on the edge of the bed. I had to pull myself together.

The phone rang, and it was a friend of mine from Perth, Phil Hardcastle. 'How are things, Brian? I suppose you're enjoying Luxor. What's the time?'

'Three in the morning,' I replied, but I was grateful for the call. Sleep was impossible. I was also grateful for the arrival of dawn.

At 5.30 am Jim arrived at the hotel in his old car. I had expected him, and was in the foyer, but that did not stop him from blowing his horn, longer than was necessary, letting everyone in the hotel know he was picking me up.

I threw my kit in the back and climbed into the front seat. 'Good morning, Captain Edwards. Did you sleep well?' he enquired cheerfully.

'Not as well as usual, Jim,' I mumbled, and we drove off at considerable speed to the airport. The night certainly had not been a good one.

Once I was through customs and immigration I would be able to get into the air and on my way. I knew I'd feel better then. We both knew customs would only be a formality; I started to feel comfortable.

Jim managed, with some shouting and pushing, to get me to the front of the rather long queue at the customs hall, and I sat contentedly on my kit while he spoke with the immigration officer. They appeared to be taking longer than usual with my passport. Obviously there was some minor problem. Jim and the immigration officer were now engaged in quite intense conversation, which was becoming more and more heated. Both men were waving their arms, and pointing at one another and then at me.

'Could I be of any help?' I enquired. My offer was not greeted with much enthusiasm, and Jim explained that my passport had been incorrectly stamped in Alexandria—I should have left the country the day before! This was really quite serious. I had been illegally in the country for 24 hours. I was to go nowhere until this was sorted out.

Slamming the door of my hotel room, I tossed my kit onto the bed. I was thoroughly dejected. This meant at least another 24 hours of sitting in my room, thinking about the problems of my following days' flying. I spent another uneasy night, worse than the night before, and there were no phone calls. I longed to speak to someone about how I felt; I wanted to speak to another pilot to talk over the navigation.

I took off for Jedda the following morning at 7 am local time. The Australian Embassy in Cairo had managed to solve the problem. Once in the air, I settled down, as usual, concentrating on the job ahead. The doubts of the preceding two nights slowly drifted away, like the featureless desert below me.

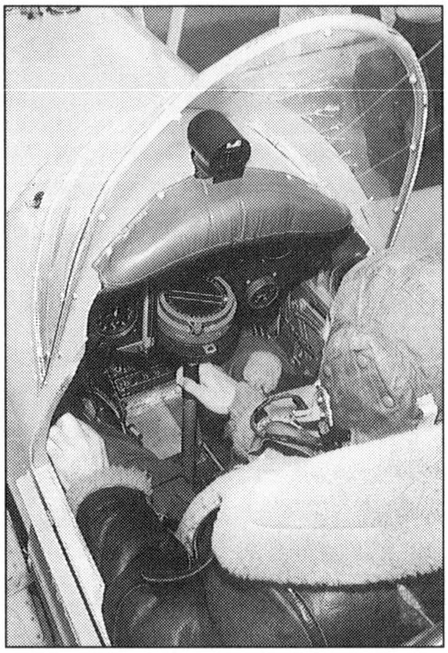

Sitting tight in Matilda.

Chapter 6

From Jedda to Ahmadabad

I

There was little to see in the desert—only the occasional unsealed road and intermittent ranges of hills and rocky outcrops. Closer to the coast there was a range of mountains with peaks up to 4500 feet, and passing over these I hit the coast, a little north of Ras Bonas. I was relieved to be able to identify this positively, and I then set out over the Red Sea on my heading. This, I estimated, would enable me to hit the Saudi coast 50 nautical miles north of Jedda.

Frequent descents to 200 feet confirmed that the wind was coming from the north-east at about ten to 15 knots. The sky was clear and the water a beautiful blue. Occasionally I would notice a ship some 2000 feet below me, and wondered if they could see my small yellow biplane.

After three hours of concentrated compass flying I crossed the Saudi coast, and when the small river system came up on my port side I knew I was on track. I flew confidently down the coast to Jedda, landing after some seven and a half hours of flying.

Saudia Airlines were of great assistance in Jedda, in fact right through Saudi Arabia. Getting through customs and immigration would have been quite impossible without their help. Unfortunately, Saudia were not able to help with air traffic control and route variations. Being without sophisticated navigational equipment and therefore unable to comply with designated airways routes in some cases, I required occasional variations to the written rule. Air Traffic Control were not prepared to arrange this. I was to find this would be the case throughout most of the Middle East.

The designated visual flight route from Jedda to Riyadh required an instrument track north-east out to the desert, for approximately 150 miles, climbing to a height of 5000 feet about 20 miles out of Jedda. I could comply with neither. The other interesting problem was the possible ground temperature of 40 to 50°C. This, of course, meant the air was less dense, and before take-off it would be equivalent to the air at a height of between 3000 and 4000 feet . My test-flying had shown that with full fuel I had a ceiling of 7500 feet. It would also take me two and a half hours to reach this height.

There was a mountain range 50 nautical miles to the east of Jedda, with peaks of between 5000 and 9000 feet. The distance from Jedda to Riyadh was 572 miles. This all added up to a possible situation whereby, if I tried to comply with the stated rule, one of the following would happen: I would get lost, run out of fuel, fly into a mountain—or all three!

An intense four-hour discussion with ATC proved useless. It was clear that if I was going to make Riyadh I would have to rearrange things a little myself.

I had spent hours the night before studying my charts, and had decided that I would have to navigate down through a small valley 60 miles to the north-east of Jedda, track around the mountains and then try to pick up what appeared to be a fairly substantial highway that went all the way across to Riyadh. This made me uncomfortable, as there appeared to be very little in the way of outstanding topographical features on the way to Riyadh. Wind and drift would have to be very accurately assessed. I couldn't help feeling that it would be like flying over the ocean, only this time over an ocean of yellow sand.

At 7 am the following day I took off from Jedda International Airport. The weather was good, but I had been told to beware of the dust storms that could develop suddenly, reducing visibility to zero.

Because of the temperature my take-off run was longer than usual. Matilda left the ground very sluggishly, and we clawed our way up to 3000 feet. I headed for the small valley, knowing it would be my only chance of getting through the mountain range.

I found the valley without much difficulty and flew down it, lifting up over the rocky ridges and skimming their flattened tops. On rising over the ridges, I would occasionally be accompanied by seemingly unperturbed, magnificent eagles who would be soaring on the updraughts. They would just nestle in next to me, often on either side, only 20 to 30 metres away. There they would stay with me, scanning the ground below, but occasionally looking across at me, and we would fly together in formation for 20 to 30 minutes—a truly wonderful experience.

The occasional herd of camels would also come into view, and once I couldn't resist flying low and scattering them. Then I saw Bedouins with rifles on their camels. I only did it once!

I arrived at Riyadh seven and a half hours after leaving Jedda. I found the road, but, as suspected, there were many others not shown on the map. By keeping a very accurate heading and allowing for drift, I ensured that everything went according to plan. I was very relieved, nevertheless, to lift over the small ridge of mountains seven hours after leaving Jedda and see Riyadh coming into view.

The only slightly irritating part about the flight was that about two hours out of Jedda my tachometer cable broke. I had to judge my engine revolutions by the engine noise and the airspeed indicator. The concern was that fuel consumption might be more than anticipated, so I worked at it all very carefully, trying to maintain what I thought was 1950 rpm and an indicated 70 knots.(The tachometer was to remain unserviceable until Dubai, despite various efforts to fix it. The Moth Club in England arranged for a replacement to be sent on a British Airways plane to Dubai. The captain proved to be a Moth enthusiast and was only too pleased to deliver it to Tony at Dubai.)

I left Riyadh the following morning for the short hop to Dhahran, only 200 miles. Easy navigation and good weather made for an uneventful flight, with the exception of being buzzed by an F16, from what I assume was the Royal Saudi Air Force.

Saudia officials met me on landing, and handled customs and immigration details very well. Before I left the airfield I spent time with ATC discussing my following day's route. To fly to Dubai, my next stop, required me to leave from Dhahran, track out over the Persian Gulf on an instrument heading and then track south to Dubai, very difficult by dead-reckoning navigation. Air Traffic Control refused to give me permission to fly close to the coast, despite intervention from the Australian Embassy in Riyadh.

Matilda was refuelled and her oil topped up. She was generally checked out and tied down—the usual ritual, ready for my early-morning departure the following day. As the bus drove me to the hotel, which was close to the airfield, the sky looked threatening, and I remember thinking that in Australia such a sky would probably mean bad weather, including thunderstorms. I thought it fairly unlikely here.

I was troubled that night. The navigation would be difficult—near impossible. I awoke at my usual 4.30 am, to an unexpected but familiar sound outside my window. It was raining! The sky was overcast; thunderclouds were rising majestically into the sky.

A gulped breakfast, and I sped to the airport. It looked reasonably clear to the south, my track direction. I quickly filed my flight plan, and was cleared by ATC to take off. There was a thunderstorm over Bahrain Airport, but it appeared to be clearing quickly to the south. Control cleared me across to Bahrain, the American-accented voice confirming that the weather south appeared to be a little better. Slowly I climbed to 2000 feet across the Gulf of Bahrain. It still looked relatively clear to the south, but it seemed to be getting darker. I flew over Bahrain and took up my heading out across the Persian Gulf. I felt very uncomfortable. The weather was bad enough, but the navigating would be worse.

The sky was getting darker, and now it didn't look all that clear to the south. Suddenly it was lit up with lightning, which was accompanied simultaneously by a deafening thunderclap. My poor old Matilda rocked violently, and I lost 500 feet in the blink of an eye. For a moment I thought the lightning had hit me. The darkness closed in like a blanket, and it began to rain. I should not have taken off.

In my desperation to get on my way, and in my concern about the route, I had really misjudged the weather. I had been entirely stupid. I was only some 40 miles out from Dhahran and 20 miles from Bahrain, but I could not reach either on the radio. I decided to get back quickly. I descended to 500 feet, turned through 180 degrees and headed back. The familiar dry mouth and moist palms appeared again.

The sky was dreadful, but at low altitude it was calm and I could see the coast quite clearly despite the heavy rain. Dhahran Control finally crackled in, and I informed them I was returning. 'I think that's a good decision, sir,' was my American friend's reply. I was very relieved to see the runway come into view, and I landed quickly.

The sky was pitch black, and the rain continued to pour down. There were more lightning flashes, but I was now on the ground. I taxied quickly to a vacant hangar, and here there was a small gathering of workmen, who were obviously following my progress with great interest. 'Better on the ground than up there, sir,' one of them said as I unhooked my harness and climbed out of the cockpit. I had to agree.

The hangar I had taxied to was owned by ARAMCO, the American-based oil company. That night I was taken out to dinner by one of their pilots, and his wife. I suspect Max and Shirley Gore took pity on me. They were delightful company, and the comments came thick and fast.

'Fancy flying this far in a Tiger Moth!'

'The weather is usually OK at this time of the year.'

'Can't understand why they insist on you flying that route to Dubai.'

'It seems impossible, really.'

We had a very pleasant dinner, and I promised to contact them when I eventually got to Australia.

The following morning the weather had cleared, and I departed Dhahran at 7 am. I was again cleared by Dhahran Control, and, as I climbed up to an altitude of 2000 feet, the American controller, who knew of the problems I had with the route, signed off with, 'Contact Bahrain on 118.5—and I would like to wish you all the very best of luck, sir!' I thanked him.

Dhahran, Saudi Arabia, April 1990.
Leaving for the second time after the thunderstorm.

Bahrain informed me that they could not keep me on radar at height below 5000 feet, and my track out over the Persian Gulf would be a dead-reckoning.

This troubled me. Flying over water was fine, but this was different. I tracked over to Bahrain and struck out into the Persian Gulf.

Suddenly, my windshield was covered in petrol, and then it immediately shattered. I called Bahrain, explaining the problem, and descended to 500 feet. The fuel had leaked out of the overflow pipe, but posed no major problems. My windshield had shattered around the base, but appeared to be holding OK.

Bahrain then informed me they had lost me on radar, and asked if operations were normal. I said they were, but I had had enough of messing around with ATC over routes.

I continued to descend to 100 feet, and scuttled back to the coast of Saudi Arabia, then took a southerly heading, clinging to the coast, moving across the Qatar Peninsula and into the Arab Emirates air space. I was very relieved to hear the English voice of the controller at Abu Dhabi, and was cleared to continue on to Dubai. It was only here that I climbed back up to 3000 feet from my previous level of 100 feet. It had been quite an eventful leg.

Tony was there to meet me. Matilda had her 50-hourly check, and a new tachometer cable fitted. I also met up here with a large contingent of

Australian pilots flying with Emirates. It was marvellous to spend a night with them. Captain Rod Cook presented me with a cheque, a donation to Legacy from all of them.

My Karachi clearance had not yet come through despite assurances that it would be ready on my arrival in Dubai. This held me up for two days, and again, as always when I was on the ground, the uncertainties began to creep into my thoughts. I desperately fought to remain positive and not think of the 'what ifs' too often. It was always very hard, however, to sit in my somewhat drab hotel room, constantly going over my maps to keep calm. A phone call to Jenny and friends always helped.

Finally the clearance came through—after a lot of 'behind-the-scenes' work by Fred Chaney Jr and Senator Gareth Evans, I believe—and I set off for Muscat, a short hop through a small mountain pass and then down the coast.

On reaching the coast, I transferred fuel up to the top tank, something I did every 15 minutes, just to make sure the top tank was always at maximum. This was a precautionary measure in the event that my fuel pump failed, although this was unlikely, as I had both an electric and a back-up hand pump. But if it did fail, I would always have a full main tank, giving me two and a half hours of flying time.

I flew slowly and comfortably down the coast to Muscat. The sky was clear, the sun out, and the navigation easy. Then I was pulled hard back to reality by the unpleasant smell of fuel.

Had I developed a fuel leak? I glanced up to the main fuel tank, to see a thin line of vaporised fuel streaming backwards over the tank. I had forgotten to turn off my electric pump, and, of course, it was pumping fuel into the already full tank. Fuel was now pouring out through the overflow pipe. I cursed myself. A lack of concentration, resulting from feeling comfortable, had caused an error that, on a longer leg, could have had disastrous consequences.

When I landed at Muscat I was told that my Karachi clearance number had come through, and everything would be fine for proceeding the following day. I was very relieved.

After taxiing over to the parking bay and turning off the engine, I realised I was still about half a mile from the terminal and it was very hot. A quick call to the tower to get permission to move closer to the hangar was cheerily received, but, having stopped the engine, I had to get out of the cockpit, prime the engine and, of course, flick the propeller. It was very hot on the tarmac, but I knew she would start easily. I undid the cowling and pressed the primer button down. The few drops of petrol came out, and I released the button. The fuel drops, however, did not stop; in fact, they increased to a steady stream! The primer button had locked down; it would not come up.

Fuel continued to pour out onto the tarmac. There was nothing I could do other than to push Matilda the half mile across the tarmac to the hangar. It was hot; she was heavy; I was tired. I was very grateful to reach the coolness of the hangar some 30 minutes later.

With the help of two technicians I worked on Matilda for four hours. Eventually we had to take the carburettor off to get the button to release. I was not keen to do this as she was running well, and there was always the danger of dirt getting in despite our being ultra-careful. But it was unavoidable, and I reluctantly agreed to the removal of the carburettor. We resolved the problem.

The following morning I departed for Karachi, one of my longest legs, 510 nautical miles over the Gulf of Amman and the Arabian Sea. The wind was calm and visibility good. However, it was completely overcast, with the cloud base hanging at about 7000 feet.

Unfortunately, as I was not able to fly anywhere near Iran, my track had to take me out over the Gulf of Amman and the Arabian Sea for 170 nautical miles before I could start to track across to the Pakistani coast. All in all, I had to fly 295 nautical miles over water. It was not my longest hop, but it did turn out to be difficult. I flew for two hours, concentrating on my heading and fuel consumption, occasionally dropping down low to 200 feet to assess the wind. There was very little, and what there was was almost on my tail.

I'm not sure exactly when I began to get disorientated, but slowly I began to feel as though the dull grey-green oily colour of the water below, which was now glassy, had merged into the grey cloud cover on the horizon. I began to think my compass was incorrect, and I was tempted to make quite ridiculous changes in heading.

I knew things were not right. I put my head out into the slipstream, held the control column between my legs and threw my arms out also into the slipstream. I sang a couple of verses of Waltzing Matilda. It all seemed strange and unreal. It was an odd feeling, one of being set apart from the aircraft. I still don't understand it, but whatever it was, I finally overcame it after about an hour.

I tried to call Muscat, but was well out of range. My transmission, however, was picked up by a Pakistani Airline 747 some 40 000 feet above me. I explained to the Captain that I was en route to Karachi from Muscat and that I had been over the water for four hours and was now about to track north for the Pakistani coast.

We exchanged much the same formalities as I had with the British

Airways captain, and once this was over we chatted for a while about my likely point of landfall. Talking to him helped a lot, and the silly feelings of the previous hour disappeared. Thirty minutes later I flew over the Pakistani coast, having picked up a small unnamed island at 20 nautical miles. It was a perfect fix and I then knew I had 200 nautical miles to Karachi. I was making good time, and the wind would now be directly on my tail.

As we crossed the coast Matilda's engine missed, just once. I think I aged ten years.

The very rugged Pakistani coast was a relief after the dull grey-green of the water. The cloud had cleared, and this also helped brighten me up. I landed at Karachi six and a half hours after leaving Muscat—this meant a ground speed of a little better than 80 knots. I was quite pleased with myself.

At Karachi my first job was to clear out the fuel tanks and the jets in the carburettor, and clean the spark plugs. Everything was fine, so the reason for the misfire over the coast remained a mystery. Some foreign material had probably got in during our removal of the carburettor in Muscat.

II

As usual, I was at the airport by five the following morning. The weather was clear and an early start looked certain. It was not a long leg to Ahmadabad, but the navigating would be difficult, as my map for this leg was basically white, with the words 'Relief Data Incomplete' in large, prominent letters across the area. I decided that to have any chance I would have to fly down the coast and over the Mouth of the Indus and pick up a fix on the Gulf of Kutch and track across to Ahmadabad.

I packed my kit into the locker behind my cockpit, checked Matilda over and called the tower for permission to start the engine.

Air Traffic Control, however, informed me that the airport director would like to see me. 'God! what have I done?' I thought. My entry into Karachi-controlled airspace had been as required. There was no problem in the circuit area, flight planning had been in order. But something was wrong. I left my kit in the plane and walked back to the control tower, to be dutifully shown to the director's office.

'Well, Captain Edwards, I've been waiting to meet you. I used to fly Tiger Moths in the 50s. Would you like a cup of tea?'

We had a marvellous half an hour, he telling me of his Tiger Moth flying, and I telling him of some of my experiences over the previous three weeks. I departed a little later than I had hoped, but I had enjoyed my talk with the director, and he came down to wave me off.

Visibility was good, and I flew down low over the mouths of the Indus, but there seemed to be hundreds of mouths and it was often difficult to pick up where the coast actually started. Here again I flew with eagles. They were not quite as magnificent as those in Saudi, but they were wonderful company, nestling in beside me.

I flew down to the Gulf of Kutch. I had decided that my fix would be the junction of three roads that appeared to meet on the coast at a small town called Mandui. It was then a track of 180 nautical miles due east to Ahmadabad.

When I reached what I thought was Mandui, the coastline proved to be nothing like that shown on the map, and I began to doubt that it was in fact Mandui.

There was no doubt, however, that I was at the Gulf of Kutch! I flew up the Gulf for some 60 miles, tracking due east, but could not make out any definite features on the coastline. It was just a maze of small rivers and mud flats that appeared to have neither beginning nor end.

I continued to fly on my easterly track, but an awful uncertainty was now starting to grip me.

The brown mudflats changed into dry land. I was due to cross a railway line ten miles after leaving the Gulf of Kutch. This did not come up, neither did the next railway line, nor the fairly large town of Jetpur. There were no fixes now for 40 miles. I was lost.

I climbed up to 5000 feet and tried to raise Ahmadabad. Nothing. I then made the mistake of leaving my heading to try to pick up a recognisable feature. But it was a fairly featureless terrain, and it was difficult to pick up the wind; however, as there wasn't any dust or sand blowing, I estimated the wind was fairly minimal. Probably my compass heading and track would have been quite similar. I flew around for about an hour desperately trying to recognise something that would enable me to get a positive fix on Ahmadabad. I couldn't get one.

I fought to keep down the panic. I just had to go about solving this predicament sensibly. I knew the relief data as stated on the map was incomplete, and the situation I found myself in was not unexpected. I had been flying for a total of five hours. I still had five hours of fuel left. Ample time to sort things out.

A couple of lakes that were on the map did not come up, but surely the fairly significant river 40 miles to the south of Ahmadabad had to be there. I changed course to head in a more southerly direction, and scanned the horizon for some indication of the river system. I found it—the unmistakable 'V' in the river where two tributaries met. I was 30 miles south of Ahmadabad, but I wasn't lost anymore.

I changed course and headed north. Ahmadabad ATC came in loud and clear. I was very relieved when the aerodrome came into view. I landed, and taxied up to the parking bay. I pulled myself, somewhat stiffly, out of the cockpit. It was hot and I was very tired. I had been in the air for six and a half hours, but the navigating difficulties had made me more tired than I would normally have been.

At customs and immigration I was shown into a very dingy, cement-floored room containing three desks, a pile of forms on each one. The piles were so high it was difficult to see the small men seated behind them, but all of the men appeared to be writing busily.

I was directed to a vacant chair situated in the front of the desk with the highest pile of forms on it. The bespectacled man behind the desk peered over at me.

'Where have you come from?' he enquired.

'Karachi.'

'Nice flight?'

I thought for a while. 'I suppose so—I'm here, anyway,' I replied.

I left the office three hours later. I don't know how many forms I filled in. I trudged, sweating, struggling with my kit, back out to Matilda, just to make sure she was OK. As I walked across the tarmac I noticed that she was at an odd angle. My plane had a flat tyre.

Flat tyres do not normally present a problem with aircraft. They are easy to change or replace. But this wouldn't be so in Ahmadabad—or so I thought. I stood looking at the tyre, trying to figure my next move, and was approached by a small man dressed in overalls.

'You have a problem?' he said.

'Yes, I'm afraid I'll have to get the tyre off and replace the tube.'

I asked him about a jack, and he said he would see what he could do. He returned about 15 minutes later with a team of three young men, and set about taking off the wheel. They appeared not to need my help, so I lay down on the tarmac with my head on the kit. It was dark by now, and the tarmac was comfortably warm. I closed my eyes and thought about the day's flying. Not all that good. The eagles that accompanied me up the Gulf of Kutch were an exciting plus; being uncertain of my position was a minus. I drifted into a wonderfully warm sleep; my body ached, and I had not realised just how tired I was. I dreamt of flying with eagles, and soaring over cliffs with them.

The friendly engineer at Ahmadabad who changed Matilda's tyre.

Soon I was awoken by a gentle nudge on my shoulder.

'I think we've fixed your tyre,' said my smiling Indian friend.

I struggled to my feet and inspected the wheel. They had done a wonderful job, and I was very grateful.

I eventually got to my hotel at 11 pm. I checked my route for the following day. Three bottles of mineral water, two Cokes and three bowls of soup later, I was fast asleep.

Although there was still a long way to go, I was relieved to be over what I thought was one of the most difficult parts of the trip. The long water and desert legs had worried me, and the flying conditions and navigation had been difficult. Now they were behind me; the period had not been without its humour!

At one airport an incident occurred that I shall never forget. I was parked below the control tower, and requested an engine start. After being given permission I started up, climbed into the cockpit and requested a taxi clearance.

'Start engine number two,' was the reply.

Being quite used to unusual happenings at that stage, I waited a respectable time and replied, 'Engine number two started, request taxi clearance.'

'Start engine number three,' came the reply.

I'm the only pilot to have flown a four-engine Tiger Moth solo from England to Australia!

Chapter 7

From Nagpur to Singapore

I left Ahmadabad early the following morning for Nagpur. The first 170 miles were again across country that was described on the map as 'Relief Data Incomplete'.

As usual I had divided my track into 20-mile segments, and concentrated on picking up the small towns and railways that did appear to be recognisable on my map. The weather was good, as was visibility, but it was with some relief that I saw my first major fix, the town of Indor, appearing over the cowling after three hours.

Taking up my heading for Nagpur, a further 200 miles to the south-east, I flew on in the crisp, cloudless sky. With the exception of the Narmada River, the country the track took me across was featureless. Later in the morning it began to get very hot. Turbulence increased, and to make matters worse I had forgotten to put my bottle of water in the cockpit.

Perhaps fatigue was taking its toll.

Soon the Narmada River came up, but the bend in the river did not have the same shape on my map. This made me uncomfortable, as it was essential that my heading should be accurate at this point: with the exception of a range of hills, the next 160 miles across to Nagpur were over difficult navigating country. I flew up and down the river for 30 minutes trying to get a position fix, but the relief on my map did not correspond accurately with what I could see beneath me.

I went back to the point where I had reached the river, took up my previous heading and continued. It was now early afternoon and the sun was very hot. Turbulence was worse, and visibility reduced because of the many bush fires dotted along my track.

I began to feel distinctly unwell. Navigational difficulties similar to those of the day before, the problem of the repairs to the undercarriage,. and my stupid mistake of forgetting to put my water bottle in the cockpit all started to take their toll. I fought desperately to maintain my concentration, knowing that relief would only come after landing at Nagpur.

A range of hills came up, but I was uncertain of my exact position although the spot heights dotted on my map appeared to indicate I was in the right position. Eventually I came over the top of the hills and down onto the plains. At this point the turbulence became very severe, and Matilda and I were thrown all over the sky, with such severity that I thought she might incur some structural damage. At one stage we were almost thrown upside down. Visibility continued to worsen due to the smoke from bush fires, and the wind had picked up. I estimated it to be 25 to 30 knots, and it had swung around onto the nose.

The last 60 miles to Nagpur were a nightmare. They took me one and a half hours. The turbulence was horrendous, and I began to develop stomach cramps. When I finally landed at Nagpur I felt exhausted, both mentally and physically. It took all the strength I could muster to get out of the plane, refuel, and get my kit out of my locker behind the cockpit. I felt light-headed and weak.

I walked slowly, lugging my kit across the tarmac to the control tower. I climbed the usual flights of stairs and literally fell into a chair. The friendly Indian ATC officer peered over the high desk and said, 'Nice flight, sir?'

I was unable to answer. My flying suit was saturated with sweat and I was covered in grease and aviation fuel from refuelling. Refuelling always tended to be a messy affair, with the fuel gushing out of the fuel tank, and with four tanks to fill I was usually drenched by the time it was over. The same thing happened with the oil. I often thought how lucky I was that I wasn't a smoker!

This time I'd made more of a mess than usual, being so tired, and I just didn't have the energy to answer the controller. I dragged myself to my feet and asked for a flight plan for the following day. I filled it out with some difficulty, enquired about the nearest hotel, and departed.

By morning I still felt a little weak, but my stomach cramps had gone and my head was clear. I arrived at ATC for a weather brief and everything appeared fine. Clear weather and a tail wind all the way to Jamshedpur. With a bit of luck I might make it to Calcutta in one hop.

But there appeared to be a minor hitch with clearances, and I was asked politely to wait in the control tower. Thirty minutes went by, and I was beginning to lose my patience. Continued offers of 'a cup of tea' only increased my annoyance. When my clearance finally came through I stormed out of the tower, muttering to myself about the incompetence on the part of ATC.

I was now one hour behind my departure time, and it was getting hotter. I was furious, and, striding over to Matilda, thumped my kit onto the tarmac and began my pre-flight inspection.

As I walked around the plane I noticed two crumpled rugs lying on the tarmac some 10 metres away. Slowly one moved, and from beneath it came a somewhat dishevelled Indian dressed in army uniform and clasping an old 303 rifle! It appeared both of these men had slept all night on the tarmac to protect Matilda. I must admit I felt somewhat chastened. I realised my impatience in the control tower had been uncalled for, and this added to my embarrassment. I thanked them both, but I doubt they could understand me. However, they did smile when I gave them a few American dollars. I was now running quite late, and hurriedly started Matilda and strapped myself in, indicating to my army friends to take the chocks away. It was then that I found I had the wrong map in the cockpit. I was furious!

Matilda-minders at Nagpur!

Motioning to my army friends to put the chocks back, I quickly unstrapped myself. I decided to leave the motor running since to restart would take another 15 minutes. Jumping out of the cockpit, I opened the back locker. I grabbed my flying bag, which was stuck, and pulled it fiercely. In doing so, I ripped the handle and the top off the bag. The contents fell onto the tarmac and were immediately whipped away by the slipstream of the propeller. Flight manuals, maps and clearance documents were all swept backwards across the tarmac. I desperately lunged toward the cockpit and turned the engine off, but not before the entire contents of my bag had been blown to some 300 metres behind the plane. I rested my head on the fuselage and swore loudly. I was now two hours late. The signs of fatigue and exhaustion should have been obvious, but I was failing to recognise them. My army friends and I spent another 30 minutes collecting all my documents and I finally took off, map in hand, three hours later than expected.

The navigation, which I had expected to be easy, did not prove to be so. I missed my first fix, and, following my rule, returned overhead to the aerodrome. I began again, and then at last I picked up my first two fixes accurately.

It was time for a drink of water. I reached down to the small box just forward of the control column in which I kept my water bottle, and realised to my horror that in my very hostile and impatient mood I had again left the water bottle in the locker behind the cockpit. God, I still had 280 miles to go before Jamshedpur! I battled on, trying to take my mind off my increasing thirst, and thankful for a rather cool, overcast sky and an increasing tail wind.

I arrived over Jamshedpur, having made good time, and was untroubled by the navigation. I thought fleetingly about continuing on to Calcutta—I certainly had enough fuel, and with such a strong tail wind I estimated it would only be another hour and a half of flying. But my thirst got the better of me, and I landed.

It was just as well I did, as I was met by members of the Jamshedpur Co-Opt Flying Club, who had been expecting me for three days. They were wonderful hosts, putting me up overnight at the Tata Steel Hostel. That night, however, my stomach cramps returned. The following morning I felt unwell: I was light-headed and had diarrhoea.

My Aero Club friends picked me up at 7 am and took me to the airport. I was feeling very ill and weary. It was only 120 miles to Calcutta, however, and I thought I could make that. I flight-planned and obtained a weather brief. Everything seemed OK, except for me!

I climbed stiffly into the cockpit; my arms and legs were aching, and my light-headedness was getting worse. I strapped myself in, fumbling to find the straps. One of the flying club members swung the propeller for me, and I prepared to taxi out. Suddenly I was racked with stomach cramps, and doubled over in the cockpit. I slowly reached over and turned off the engine. I glanced over at my Flying Club friends, who were looking on anxiously.

'I don't think I can fly today,' I mumbled.

The rest of the morning is difficult to recall. I vaguely remember being driven back to the hostel and being visited by a young Indian doctor, who left a large number of tablets and prescribed rest and fluids.

I slept for 24 hours, and awoke clear-headed and feeling strong. A combination of fatigue, dehydration and possibly some unusual food had been the cause of my problem. I had missed the early signs, which, in retrospect, were quite obvious, and I made a mental note not to let this happen again. I still had a very long way to go before I reached home.

At Calcutta I was met by Tony. It was good to see him, and he set about checking Matilda out, preparing her for the long haul down to Singapore, where he would do a 100-hourly check. Everything seemed OK. From my point of view she was handling well, although I had been concerned about the battering she had taken from the turbulence we experienced in India. Tony assured me that she was sound.

The two days' rest in Calcutta was much needed. A couple of long phone calls home to Jenny and my family lifted my flagging spirits, as did phone calls from friends. Len Findlay from the West Australian and Gerry Gannon from the ABC also kept in touch with me, and I really appreciated their support.

While in Calcutta, Tony and I were lucky enough to be given an audience with Mother Teresa. I have since been told that it is quite difficult to obtain an audience with this magnificent woman, but I must confess all I did was hop in a taxi and ask to be taken to see her.

The drive through the narrow streets of Calcutta was quite an unnerving experience. The sight of people lying in the gutters begging was terrible, far worse than I had imagined. We pulled up at a small alleyway and my taxi driver, leaning back over the seat, said, 'This is the hospital. Go down there and knock on the door.'

Tony and I walked down the alleyway to a large wooden door, in front of which a small crowd of people were standing. I pushed through the crowd and knocked. The door was opened by a young, fresh-faced nun. She asked what I wanted. I replied, 'I'd like to see Mother Teresa.'

She ushered Tony and me in and asked us to wait. After a few minutes, from behind a curtained doorway came a tiny, frail lady. She was dressed in a simple grey habit with a blue veil. She put out her hand. 'Welcome to Calcutta—where are you from?'

'I'm on my way to Australia from England.' We chatted for a while, and then went across to the hospital. We were taken into a very clean ward, where about a dozen quite severely incapacitated spastic children were being cared for. The standard of treatment and management was obviously very high, and I must admit I was surprised. Mother Teresa explained that they had many helpers from all over the world. People of all ages and nationalities came on a voluntary basis, spending time with her and helping out.

'Mother, what will you do after retirement at the end of the year?' I asked.

She replied, 'Oh my dear, I won't have a spare minute—I have so much praying to do.'

To me this was an extraordinary answer. It had never occurred to me that someone might have difficulty fitting things in because of prayer, although I must admit I had been doing a little praying myself on occasion over the past three weeks. Meeting this woman was truly inspirational.

I left Calcutta on Sunday, April 22 after being held up for an hour because of bad visibility. I was cleared to fly at 2000 feet across to Chittagong—a leg of only 200 nautical miles. As I flew over the India–Bangladesh border, it began to rain and the cloud base dropped to 1000 feet. I scuttled along under this, but flew most of the time in fairly steady rain.

The mouths of the Ganges, of which there are thousands, stretched endlessly below. All I saw were dark green foliage, dull green water and a grey sky. The only good part about this leg was that there appeared to be little wind and consequently no drift, and I was able to get an accurate fix before the short hop of 40 miles across the Bay of Bengal to Chittagong.

The constant light rain made visibility a little difficult. As with the mouths of the Indus, the mouths of the Ganges did not match those shown on the map I was holding. I picked up the southern tip of South Hamia Island and eventually crossed the 40 miles of water to Chittagong. I couldn't help thinking that it was here (or rather, a few miles south, across the Bay of Bengal) that, 60 years earlier, trying to break the record from Australia to England, Sir Charles Kingsford Smith was lost in bad weather. The grey waters of the Bay of Bengal below suddenly looked terribly inhospitable.

Chittagong airport appeared, and I was cleared to join the circuit and land. As usual the wind sock was horizontal, and, of course, blowing directly across the strip. The cross-wind landing was awkward, and I had a lot of difficulty keeping Matilda straight and on the runway.

The wind was too strong to taxi, and I called the tower requesting 'wing-tip assistance', but this obviously caused considerable confusion and I was asked repeatedly to 'say again'. The confusion increased, as did the wind, but fortunately I had kept the engine going and held Matilda into the wind, afraid to turn in case the wind lifted me over onto one wing. I was in this position for some 15 minutes, desperately trying to hold the plane straight and endeavouring to explain to the tower why I was unable to move.

Eventually, three fire engines and an ambulance came down the runway at great speed, sirens sounding and red lights flashing. They screeched to a halt at a safe distance from my plane. Out jumped half a dozen personnel, who all rushed up to the cockpit. Somewhat embarrassed, I explained, for what seemed like the thousandth time, that all I needed was for someone to hold the wings while I taxied. Fortunately, everyone was very good-natured about the misunderstanding, and I was escorted to the parking apron by nurses, firemen and army—quite an entourage!

The usual ritual—refuelling, tying Matilda down, getting through customs and immigration, flight planning—was carried out, and four hours later I arrived at my hotel. I spread my books out, and began to study the following day's leg to Rangoon, which would be 400 nautical miles—not too long. The navigation would be fairly straightforward, as it was coastal most of the way and I had chosen Gwa as my last coastal fix—this was marked on the map as a minor airstrip. From there it was only 100 nautical miles across to Rangoon. I was to be very grateful that I chose Gwa.

Everything was in order. I packed away my maps and stretched out on the bed. I made a quick phone call home to Jenny and the family. Time to go to sleep.

Suddenly I realised I had run out of American dollars, as I had used my last when refuelling at the airport, and I had been told that credit cards would be no good in Burma. The Department of Foreign Affairs had already explained to me that there had been a lot of difficulty in obtaining a clearance to land at Rangoon, and I must only stay one night as there was considerable civil unrest.

I jumped up, grabbed my wallet and rushed down to the hotel lobby. There was an American Express office close to the hotel, and I was accompanied there by the young hotel bellboy. I must have looked quite a sight as I walked into the office. I was still dressed in my somewhat grubby flying suit and extremely dirty boots, and my hands were still covered in grease from the refuelling. I was told by the teller that there was no way I could obtain any money on my card. End of round one. I asked to see the manager, and after some arguing was directed into a very well-appointed office. The polite, very well-dressed Bangladeshi behind the desk invited me to sit down.

I hesitated, as the chair was obviously a very expensive and clean one, and I felt it would never be the same if a grease-covered individual such as I sat in it. The Bangladeshi gestured, and, on seeing my hesitation, said, 'Please be seated. Now what can I do for you?'

'I need 500 US dollars,' I said.

'Why?'

'Well, I'm on my way to Perth, Western Australia, and I'm flat broke.'

'Well, that's OK. Let's have your card. Yes, I see there should be no problem—it should be through in a week.'

'Oh—that's no good,' I replied. 'I need it this afternoon.'

'You mean you need US dollars this afternoon ? You realise you are in Chittagong, Bangladesh and it's Sunday?' he asked, somewhat incredulously.

'Yes, I do—it's awkward, isn't it?' I replied.

We sat looking at one another across the desk.

'Come back in one hour—I'll see what I can do.'

One hour later he handed me 500 US dollars. I thanked him and he shook my hand.

'I certainly hope you make it to Perth,' was his only comment.

The following day I was out at the airport at 6 am. The weather looked good, and after completing departure formalities I started Matilda, strapped myself into the cockpit, and taxied out to the holding point.

Always at this time I became a little nervous. Did I have everything prepared for the flight? Would the weather be manageable for the whole leg? My navigation had to remain accurate, my time intervals precise; I had to have my bottle of water in the cockpit—everything had to be right.

'Chittagong tower: Hotel Papa Hotel ready.'

'Hotel Papa Hotel: Chittagong tower. Hold position.'

'Oh, come on,' I thought. 'Let me get going—there's no traffic, and the weather always deteriorates in the afternoon. I need to be well on my way by then.'

Five minutes went by. A Gypsy Major engine does not like idling for too long—the plugs foul up, causing a magneto drop, and this often requires stopping the engine, cleaning all the plugs and starting again. This usually takes about an hour, if you're slick!

I called the tower again, and advised them that I was ready. I requested immediate take-off.

'Hotel Papa Hotel,' I heard, 'return to parking bay. Your Rangoon clearance has been cancelled.'

'What!—say again!' I spluttered over the radio.

'Hotel Papa Hotel, Chittagong tower: return to parking bay. Your Rangoon clearance has been cancelled.'

I swore profusely (not over the radio), and taxied somewhat faster than I should have over to the parking bay, switched off the engine, jumped out of the cockpit, ran up the control tower steps and stormed into the operations room.

'What do you mean, my clearance is cancelled?' I shouted at the now wide-eyed air traffic controller.

'I'm sorry, sir, but—'

'You're sorry—listen, I'm the bloody pilot and I've got to sit on my backside here. I need to be on my way.'

I must admit the air traffic controller showed immense patience, and finally it was established that for some inexplicable reason Rangoon had decided to cancel my clearance. I left the control tower dejectedly, and slowly walked back to my plane. There was nothing I could do. I might be here for a week!

I hauled my kit out of the locker behind the cockpit and threw it under the wing, climbed under it myself and rested my head on my bag. I would just have to wait.

About two hours later an official came towards me across the tarmac. I watched him walk over, but decided I would remain lying down under my wing—damn them!

He knelt down and said, 'Captain Edwards, your clearance is through. You may depart.'

Fifteen minutes later I was on my way down the coast, heading for the Bangladesh–Burma border. It was now late morning, and I was four hours behind schedule, but the wind appeared to be favourably on my tail and I hoped I could make up a little time.

Trouble struck when I had been flying for some five hours. Everything had been going well. The wind had swung around on to the nose and my ground speed had dropped a little, but I was on track and the navigating had been relatively straightforward. I flew over a small strip—'Sandoway', a good position fix. I now had only 160 nautical miles more.

The weather had remained reasonably good. The coast was attractive: the jungle came right down to the beautifully white beaches, against which an emerald-green sea lapped. Despite the troubled beginning, this leg was becoming quite a pleasant experience.

Then I detected a slight change in the sound of the engine. At first I ignored it, but there was definitely something different there, a slight fluttering sound. It increased, and I began to lose power. There was no doubt about it, something was wrong. The engine power continued to fall, and the fluttering noise became louder. Obviously I would have to land. I estimated that I was only five miles from Gwa, and although it was marked as a minor airfield it would be better than putting down on the beach.

Within six minutes Gwa came up. Engine revolutions were now falling alarmingly, and the noise was much louder. I decided I would have to put her down immediately, and dispensed with the usual low-level run to check the state of the airfield. The wind was 15 to 20 knots, coming straight off the ocean and blowing directly across the airfield. The first 100 metres of the gravel strip looked reasonable, but I wasn't certain about the rest, so the landing would have to be very accurate.

I landed almost without power, the RPM having continued to drop. I checked the magnetos, which seemed all right, and turned off the engine.

I had seen a small village near the strip as I flew over, but I hadn't taken much notice. I was more concerned with my loss of engine power. But then, out of the surrounding jungle came a whole crowd of smiling, brown-skinned people. They surrounded my aeroplane, laughing, shouting and jumping up and down. Initially I was very concerned. They certainly seemed friendly enough, but I thought they might damage Matilda.

I quickly undid my harness, and stood up on the cockpit seat. I threw my arms in the air and shouted, 'You will all stand back!'—and they did! No one was more surprised than I was. Some 100 or so people now formed a wide circle around Matilda and me.

I jumped out of the cockpit and ran around to the front of the plane. I immediately spotted the problem. Some protective tape placed on the leading edge of the propeller by the manufacturers had come loose and wrapped itself around the blades, pulling the propeller out of pitch. This accounted for the drop in engine power and the fluttering noise.

I quickly removed the tape, and, waving goodbye to my friends, was back in the air in five minutes. Two hours later I landed at Rangoon International.

After the usual three hours of refuelling, customs and immigration, I reached my hotel and fell exhausted onto the bed. It had been another difficult day. I lay for a while going over the day's events, and eventually hauled myself up and ordered soup and began preparing my maps for the following day. Two hours later I was in a deep sleep.

At 3 am I was awakened by a ringing telephone. I fumbled for it in the dark, knocking it onto the floor. Finally, still half asleep, I mumbled, 'Hello?'

'Captain Edwards, this is the Myanmar police. We believe you made an unscheduled landing on our coast today. What was the reason?'

I had forgotten to mention my stop after landing. My sleep-garbled explanation didn't seem to make sense, certainly not to the police. They suggested I meet with them before leaving.

Later in the morning at the airport I was able to explain satisfactorily to them the reason for my unscheduled landing. Mid-morning I took off for Bangkok, pleased to be on my way.

This was not to be a long leg, only 320 miles, but it was over the line of mountains that run down the peninsula virtually all the way to Singapore. It meant climbing up to 7500 feet, but I knew that I would have used up some two and a half hours of fuel by the time this was necessary, and my rate of climb would be considerably better because of this.

I flew out across the Gulf of Martabar over Moulmein, and then began to climb towards the mountains. The country was spectacular: towering mountain outcrops, surrounded by dense tropical rainforest. It was the only leg over land, however, on which I stopped looking for a place to put down in case of an emergency—there was nowhere to land. Had something gone seriously wrong, I would have had to put down on top of the trees. It was certainly with some relief that I finally flew down onto the Thailand plains the other side of the mountains and landed at Bangkok International.

I walked from my parked plane towards the large terminal. I was elated to have finished what I thought were probably the most difficult parts of the whole flight, but was not looking forward to the usual three to four hours of customs and immigration checks. Lugging my kit, I entered the terminal, and stood looking for a sign indicating customs.

As usual, I was approached by a police officer, only this time he was adorned with the most extraordinary array of gold braid and medals and was wearing not one, but two, extremely large hand guns.

'Here we go again,' I thought, and braced myself for the usual barrage of questions, the unbelieving looks once I answered them and, finally, the officious methods of the customs and immigration officers.

This officer, however, walked towards me, held out his hand in welcome and said enquiringly, 'Concorde?'

I turned around, thinking there was someone behind me, but there wasn't. Here I was, standing in my flying suit, with four stripes on each shoulder, presumably looking like a Concorde Captain! Well, why not!

'Yes,' I replied, but even I didn't think that it sounded all that convincing.

He vigorously shook my hand, called someone over to carry my kit, and took me through customs and immigration in the twinkling of an eye—no questions, no inspection. Once we were in the main terminal, he again shook my hand vigorously and passed me my kit. I thanked him, and said I would take him for a ride one day. We parted on the best of terms!

I stood in the centre of the busy terminal, a little uncertain as to where to go. I had never had such an easy transit through customs. Suddenly I was thumped heavily on the shoulder, the force of the blow sending me sprawling over my kit. I tried desperately to regain my feet—how undignified for a Concorde captain—and, turning, I saw the smiling face and large, bulky frame of Joe DePledge.

'Eddy, you hopeless bastard—where have you been?' he asked, holding out his hand.

Joe was a very good friend of mine, and a fine pilot too. He had flown up from his station, Mandora, in Western Australia's Pilbara region, to give me a little moral support. It was marvellous to talk to him and relate some of my experiences—some good, some not so good.

With Joe DePledge in Singapore.

While in Bangkok I also received another telephone call from Gerry Gannon of the ABC. Gerry had been following my progress all along, and it was always a delight to talk to him. I remember him asking me how Matilda was holding out, and I replied that she was looking a bit old, battered and tired—just like the pilot. But I was beginning to feel as though I were getting a little closer to home.

One day's rest in Bangkok and I took off for Phuket—weather good, navigation easy. I flew along, marvelling at the beautiful coastline, with its little villages, dark-green jungle, and small river system. Apart from being shadowed at one stage by a Thai Air Force Pilatus, it was a thoroughly enjoyable flight. As I flew over Surathani, I was called up by the tower. 'We like your aeroplane. Would you like to land?'

Unfortunately I felt I had to keep going, and I explained this to them. Their disappointment was obvious, but they wished me well and I flew on to Phuket, arriving in the afternoon.

The following morning I departed for Singapore—510 nautical miles, but a fairly straightforward leg, or so I thought. It was not to be the case.

The forecast for most of the leg was good: wind calm and visibility excellent. It was a beautiful morning, and as I climbed out over the small islands that lie south of Phuket I began to think just a little about arriving home in Australia. It was a luxury I had not allowed myself at any stage of the trip, but I was getting close now. Next stop Singapore. Tony was to do a 100-hourly check and then soon I would be home to family and friends.

The coastline was magnificent, the navigation easy, and the weather good. I passed over the Thailand–Malaysia border and flew down towards Kuala Lumpur, planning to overfly Kuala Lumpur International. Originally the plan had been to have an overnight stop here, but, with the weather being so good, I thought I could easily make it on to Singapore.

At this stage the wind had picked up, and my ground speed had dropped to 55 knots. It was a little slower than I had anticipated, but there was no need to worry. Over Kuala Lumpur the traffic was heavy, and I was directed quite a way off course and then back to the coast. This took some time, and when I was on track again I realised that I had been flying for six hours and still had 180 miles to go.

I began to feel just a bit uneasy as I continued down the coast. I realised it would have been far more sensible to have landed at Kuala Lumpur. The wind continued to pick up, and was on the nose. My ground speed appeared to be falling. Another hour and a half of flying, and I passed over a small coastal town, Muar. I still had 100 miles to go to Singapore with probably two hours of fuel left, and although the wind seemed to have eased I began to worry about making it.

I quickly made the decision to try to land somewhere and get fuel—even if only from the local garage. I had seen a small airfield a few miles inland from Muar, and I headed for it now. It was nestled in the jungle, but I found it fairly easily and the gravel strip seemed in good condition. I swung Matilda around and gently touched down, taxiing to the end of the strip.

There were a few what appeared to be army huts on the perimeter, but no one was in sight. There will be a garage nearby, I thought; just a few gallons and I will be on my way. I stiffly hauled myself out of the cockpit, realising I had been in the air for eight hours, quite a bit longer than anticipated.

At that stage I noticed a fellow walking quickly across the field towards me. He was dressed in civilian clothes and appeared to be very concerned. As he approached me I held out my hand.

'My name is Edwards. Sorry to drop in like this, but I'm just about out of fuel. All I need is a few gallons from the local garage and I'll be on my way to Singapore.'

I didn't think that he understood me completely, but I was used to this.

'Do you know where you are?' he asked.

'Of course, Malaysia.'

'Yes, but do you realise this is a high-security paramilitary base?'

I said that of course I didn't know that, but I would be on my way in 30 minutes.

He gave me an odd look, and then a car came roaring across from the perimeter of the base. Out stepped two young men in uniform—the local police. Obviously I was not going to be on my way in 30 minutes.

I was escorted to the local police station and ushered into a small room with one desk and two chairs. The large, relaxed superintendent behind the desk was a stark contrast to the two young, rather agitated officers who had picked me up. I was interrogated for three hours, sometimes fairly intensely: what was I doing in Malaysia? Why had I landed at a high-security base? When it was finally explained to their satisfaction, we all left, returning to the airfield for an inspection of the aircraft.

The size of the aircraft did not fit in with my explanation of being en route from England to Australia, and the questioning began in earnest again. After a further hour my interrogators appeared to be relatively satisfied, and I was finally taken aside by the chief.

'OK, Captain Edwards, there are only two things in Malaysia punishable by death; one is carrying drugs and the other is carrying firearms. We are happy about the former now, but how about the latter?'

I mumbled that of course I didn't carry a firearm.

'Fine,' he replied. 'It is too late for you to continue now. You can stay here overnight and leave tomorrow.'

I must admit I was rather relieved. I thanked him, and was invited to stay on the base with the senior officer and his family. It was a very pleasant night, and I departed early next morning, stopping at Johor Baru for customs formalities before proceeding to Singapore.

On reflection, of course, I realise the Malaysian authorities acted quite correctly. However, what had begun as a simple leg turned out to be rather awkward. But I don't think my little yellow Matilda looked like a Stealth Bomber or a secret weapon!

I arrived at Singapore early the following morning, and I was met by Tony and Joe. Tony carried out the 100-hourly inspection during the day, and we spent a pleasant night with Greg and Juliana Harvey. Greg was the

Singapore Manager of British Petroleum, and an old schoolfriend of mine.

As far as Matilda was concerned Tony was happy with everything and, with the exception of having to have a tube in one of her tyres changed, she was ready for the last stretch to Australia.

I might add, however, that changing a tube in a Tiger Moth tyre is quite difficult. Joe, Greg and I sweated for three hours at a local garage completing what we all thought would be a simple task!

Senior officer and family at the military base in Malaysia where I stayed overnight.

Chapter 8

From Palembang to Kupang

I departed early on 30 April for Palembang. The navigation would be fairly easy—some rain and low cloud close to Palembang, but not much to worry about. I was farewelled by Tony, Greg and Joe, and took off and climbed out over the Singapore Straits, picking up the small islands and heading for the coast of Sumatra.

I crossed the mangrove-covered Sumatra coast on track, and turned to a southerly heading across the Berhala Strait towards Palembang. The weather began to deteriorate as expected, with cloud base coming down to 2000 feet with rainshowers, but visibility remained fairly good. I continued down the coast, picking up the large river complex there, and tracked without difficulty through the grey murk across to Palembang.

On landing, I was greeted by a group of excited airport officials, who made me feel very welcome. I had flown through fairly constant rain, and everything in the cockpit, including me, was soaking wet. I quickly went through the usual airfield formalities—customs, immigration, refuelling—and then hurriedly left for the hotel.

At the hotel I set about drying maps and clothing, and after three hours, a call home and a bowl of soup I settled back on the bed for a sleep. I lay there looking out of the window and drifted contentedly into that pleasant, dreamy state experienced before deep sleep. Having been constantly on the watch for weather changes, I continued to look hazily outside at the sky. It was getting darker, much more quickly than usual.

I forced myself to get up and take a proper look at the sky. To the south-west the sky was purple. Huge cumulonimbus clouds were building up. Forked lightning was crackling down to the horizon. The storm would be on us within 30 minutes.

'Well, so what?' I thought dreamily. 'I'm on the ground', and I dropped back on the bed. Suddenly, horrified, I realised that Matilda was parked out on the tarmac. This storm would tear her apart.

I jumped off the bed, hauled on my flying suit and ran out of the door, only then realising that I had no shoes on! I ran back, grabbed my boots, and rushed downstairs and out into the street.

The sky was becoming darker, a terrible purple colour. The wind was picking up and blowing paper along the footpath and gutters. A taxi appeared.

'To the airport!' I shouted as I flung myself into the back seat, 'I have to get there before this storm!'

I'm not sure the taxi driver understood everything I said, but he certainly understood the urgency in my voice. We drove at breakneck speed to the airport, which was some 20km away. The storm literally chased us, following 2 to 3km behind.

When we reached the airport, I jumped out of the car and ran through the crowded terminal, straight past the security guards and out onto the tarmac. I could hear people shouting excitedly behind me, but I just kept going.

Matilda was parked not far from the terminal, and I sprinted across the grass to her. The storm was just coming over the perimeter of the airfield. I desperately pushed her toward the only hangar I could see, 400 metres away. The people in the terminal were all now pressed up against the windows, watching me excitedly. They seemed to be enjoying the performance very much.

I got to the hangar as the storm broke over the airport. There was a truck in the hangar, and a dozen or so men were taking shelter. I shouted to them to get the truck out. They didn't seem to understand, so I ran in to move the truck myself. Pushing aside the astonished workmen, I pulled the cabin door of the truck open and leaped inside. Fortunately, the keys were in the ignition. The truck engine roared into life, and I drove frantically through the hangar doors and sped out onto the tarmac, scattering the workmen, who were now looking on with a mixture of disbelief and panic! Then I pushed Matilda into the hangar.

The astonished workmen just stood in the corner, speechless. I climbed out of the truck and tried to explain that I had really had no alternative. I stood looking from my protected position in the old corrugated iron shed as the storm roared across the airfield—a combination of rain, thunder, lightning and gale-force winds. Matilda would have been destroyed had she been out in it.

It was all over after about ten minutes, and relative calm returned. I decided to leave Matilda in the shed, and managed to squeeze the truck in as well. Nobody seemed to mind.

I got back to the hotel about three hours later, absolutely exhausted.

My plan for a comfortable early night had been dashed. I went over my route for the following day, and fell into bed.

The following morning I departed early for Djakarta. It was overcast, with a grey mist of drizzle occasionally sweeping across the airfield, but the weather was forecast to improve on track.

I took off, and headed out in a south-easterly direction. As usual, the weather deteriorated, and I was forced to scuttle along under the cloud base, which was down to 800 feet, trying to dodge the heavier showers. Often this was impossible, and I was soon soaking wet. This inconvenience, however, was more than compensated for by the thrill of flying low over the jungle. When I passed over small villages the people would come running out of their huts and wave frantically. I would reply by waggling my wings, and occasionally doing a low orbit. It was a wonderful experience.

On arrival in Djakarta, after four hours of flying, I was met by the Australian Air Attaché from the Department of Foreign Affairs, Group Captain John Bushell, the General Manager of Paragon Resources, one of my major sponsors, and members of the local press. It was a very warm welcome, and I was invited to spend the night at the Group Captain's house. I was very grateful for his kindness and hospitality.

The next two legs to Surabaja and Bali involved fairly straightforward navigation, and the weather was good. For the first time since I arrived in Indonesia it didn't rain, and I arrived dry on each occasion.

At Bali I was met by John Elliot, the Australian Consul, and was invited to dinner at the Australian Consulate. It was a pleasant night, but I found it very hard to relax knowing that the next day I would have only two more legs to go before arriving in Australia. I had also developed a small leak in the top fuel tank, and that was worrying me despite advice from a helpful Qantas engineer that my makeshift repairs would hold until Perth.

The departure from Bali for Waingapu, on the small Indonesian island of Sumba, was not without problems. I had arranged for fuel to be available at the Waingapu airport, but as I was about to leave I was informed that no fuel was available. This annoyed me immensely, as I had spent some hours the day before sorting this out and had been assured that everything would be in order. After a further three hours of intense discussion, I was finally advised that I would have to find the nearest garage on arrival!

I departed Bali at 11 am and flew east over the beautiful Indonesian islands. The sky was clear and the air crisp. As I flew along the southern coast of Lombok Island the towering, silent volcano Rindjani came into view. It was a magnificent sight. This, coupled with the beautiful white beaches that separated the dense green jungle and turquoise sea, moved me to song: just a few numbers from Rogers and Hammerstein's 'South Pacific'!

A short 70 miles across the Sumba Strait the coast of Sumba came up. Soon Waingapu airfield, nestled into the headland of Waingapu Bay, came into view.

On landing, I set about trying to find a garage. As it turned out, there was only one on the island, and after hiring a small utility and a 44-gallon drum I proceeded to the garage and returned to the airfield with 40 gallons of ordinary 'super' petrol. The refuelling was a precarious affair; I stood on the fuselage, being handed up buckets of fuel, which I was transferring into the tank after straining it through a chamois. After two hours I had transferred 120 litres to my ferry tanks.

I still had 90 litres of clean aviation fuel in my top tank and one of the ferry tanks, so I decided to not put any more of the 'super' petrol into the fuel tanks. The fuel from the garage was quite dirty, as shown by the residue on the chamois. As the next leg to Kupang was only 200 miles, I would now certainly have enough.

The following morning, for the first time on the whole trip, I managed to get away on time, 7 am. I started out across the Savu Sea towards Kupang. After one hour of flying, I realised that the headwind was a little stronger than forecast, and although the total distance to Kupang was not far, the now familiar and uncomfortable feeling of doubt slowly began to creep into my thoughts.

A small island some 100 nautical miles from Waingapu was my fix, but after one and a half hours of flying I could not see it. I had begun to transfer fuel up to the top tank, and felt more and more uncomfortable—both about the dirty fuel and about whether I had enough.

I decided to return to Waingapu and fill up completely. Dirty fuel was probably better than no fuel at all.

I left Waingapu for the second time at midday, with full tanks—as usual, later than expected—but I flew without incident to Kupang. As it turned out, the extra fuel was not needed, and I had probably been a little overcautious. However, flying over the water is difficult enough when navigating by dead-reckoning, and worrying about sufficient fuel just adds to the problem. I was not sorry I had turned back.

Landing at Kupang, I emptied all of the automobile fuel out of the tanks, gave it to the locals, and refuelled with clean aviation fuel. Cleaning out my filter lines took a couple of hours, but finally I was satisfied no dirty fuel was left in the system. The locals were very happy to have the fuel, and I was relieved to have it out of my tanks. I gather some of them went off and sold it. Private enterprise at work, I suppose.

The afternoon was spent tinkering with Matilda, checking everything I could. I had started to think now about getting home—only 350 miles.

I knew that Jenny was going to be at Kalumburu, an Aboriginal mission in the far-north Kimberley region of Western Australia, and Michael, Christopher and Susie would be waiting for me at Broome. Thoughts of making it home to Australia consumed me.

Just one more leg to go, 350 miles. The fact that 300 miles was over water was of no consequence. I tied Matilda down for what I thought would be the last night out of Australia. Everything was ready for the last hop home.

Matilda's cockpit—getting smaller as time went by!

Chapter 9

From Kupang to Kalumburu—to Perth!

I walked out of the small Kupang terminal and climbed into the first taxi. 'The nearest hotel,' I shouted cheerily to the driver. I would soon be home.

I now found it difficult not to think about arriving home in Australia. One more leg, 350 miles. Of course, there would be another 1000 miles to Perth, but this seemed insignificant. At least I would be home!

Before going to bed, I walked outside to check the sky—something I had done every night for six weeks, although this time I felt a little more anxious than usual.

This time tomorrow night I would be home. My first landing in Australia would be Kalumburu. Fortunately, Australian Customs had agreed to let me do this, and were to meet me on arrival in Broome. Jenny would be at Kalumburu.

The sky looked clear and the stars were out; it seemed as though conditions would be favourable for my last leg into Australia.

I awoke at 5 am, and walked outside to check the sky once again. The wind was calm and the sky was clear, but over to the east I saw great towering cumulonimbus clouds that made my heart sink. I asked my driver on the way out to the airport what he thought the weather was going to be like.

'Lovely,' he replied, 'it's always lovely here in Kupang—it never rains.' With that he turned on his windscreen wipers.

At the airport it became obvious I would not be able to fly. A storm was approaching from the east; the wind had picked up and the rain was steady. The bad weather en route was confirmed by Darwin Control, and a note from the senior controller in Darwin suggested that it would probably be better the next day.

Resigning myself to another 24 hours in Kupang, I spent the day tinkering with Matilda and talking to pilots at the airport. I sat outside the hotel that night looking at the sky. To the east lightning persisted, as did huge cumulonimbus clouds. It didn't look good ...

The following day the weather was the same—no chance of flying. I moped around the airport again for a while, and then my taxi driver, 'Alpho', who had befriended me, suggested he show me around Kupang. Alpho was a rather distinguished-looking man: collar and tie, neatly pressed pants; these were a little out of place under the circumstances, I couldn't help feeling. Perhaps I would like to have lunch with him and his family? I was very grateful for this as my frustration was now building up. I rang Kalumburu Mission and spoke to Jenny, who did her best to lift my spirits, telling me that the many people who had been following the flight would not mind waiting another day and that Michael, Christopher and Susie were at Mandora Station and would be waiting at Broome to meet me there. Then my friend and I left to see the sights of Kupang.

We drove around Kupang, and after a while we stopped at what he called an 'Australian Bar' for a cold beer—a rather dingy bar just off the main street in Kupang, close to the beach. It was filled with loud-mouthed, overweight, drunken Australians. The raucous laughter was interrupted by the squeals and giggles of the local women, intermittently being pawed by the men.

My friend and I ordered a beer at the bar and I could see the hurt in his eyes as he watched the young girls of his country dangling themselves around the drinkers at the bar. I felt equally embarrassed at the actions of my fellow countrymen.

We left our drinks half-finished and departed, both feeling glad to be out of the place. He took me back to his home, a small, simple house close to a busy street. I met his wife, two daughters and young son, who all welcomed me warmly, and we had a delightful lunch.

Returning to my hotel early in the evening, I again sat outside looking at the sky. The storms still persisted to the east. Maybe the weather was a little better than the day before, but it still did not look all that good. Later that night I worked out a fairly simple method of navigating around the intermittent storms that I thought I might encounter across the Timor Sea. As I only had a compass and was flying over water where visual fixes were not possible, it was important to have a fairly simple system that would enable me to leave my track, go around the storm and return to my original track.

I used a simple system that involved flying into wind on a specific track, and for a set time. After a couple of hours I was satisfied with my figures, which I would adjust the following day after obtaining wind direction and speed.

Alpho picked me up early the following morning and we drove to the airport. The weather was a little better, but it was with despair that I read the weather report from Darwin. The same as yesterday, and it was probably going to stay that way for a few days.

I walked out of the control tower building, and my friend could see I was disappointed. 'Not going home today, Brian?'

'I'm afraid not.'

He offered me a cigarette, which I took. I sat down on an old gun emplacement, a relic from World War II, and smoked my cigarette. I hadn't smoked for 25 years, and I really can't remember what it tasted like. We went back to the airport terminal and sat drinking coffee.

I spent more time fine-tuning my system of dead-reckoning, and continued to adjust my figures in relation to the wind speed and direction in the weather report. It was a useful exercise. I knew the information given me from Darwin would be accurate and I felt I could use this information with confidence. My intended track was 143 degrees, and this was to bring me over the Australian coast 50 miles to the north-east of Kalumburu. The wind was forecast at 120 degrees at 25 knots. To maintain this track I would have to steer 135 degrees, and I would have a ground speed of 46 knots.

I decided that, as I came up to the intermittent storms predicted for my track, I would head directly into the wind, which would give me a ground speed of 45 knots and a track of 120 degrees. If I held that for 30 minutes I would have gone 22.5 nautical miles and be approximately five nautical miles off track. If I then tracked 180 degrees (heading 162 degrees) I would be back on track after about 17 minutes. On the other hand, in the unlikely event that I would have to go to the right to get around the storm, I would again track 180 degrees for 15 minutes and then 120 degrees for 32 minutes. The diagram below illustrates this. Providing the wind remained constant in direction and speed, I needed only to remember my headings and time intervals, and of course I would have the diagram strapped to my knee.

Alpho sat quietly, and I was grateful for his company as I went over my navigational system for crossing the Timor Sea in the current weather. As is often the case, the local weather was good, but I knew it would deteriorate over the ocean.

After about an hour I received a report, from the pilot of a helicopter flying near my intended track. He reported that the weather was clear, and ATC asked what my intention was. I decided I would go.

Alpho held out his hand. 'I think the time is right for you to go, Brian. Good luck.'

I shook his hand; we had struck up quite a firm friendship. 'Maybe we will catch up again, Alpho. Thanks for everything,' I replied, and turned and ran out to Matilda, sitting expectantly on the grass near the strip. She started with one flick of the propeller. I climbed into the cockpit and taxied out. It was difficult to contain my excitement. I was on my way home. Only one leg of 350 miles remained.

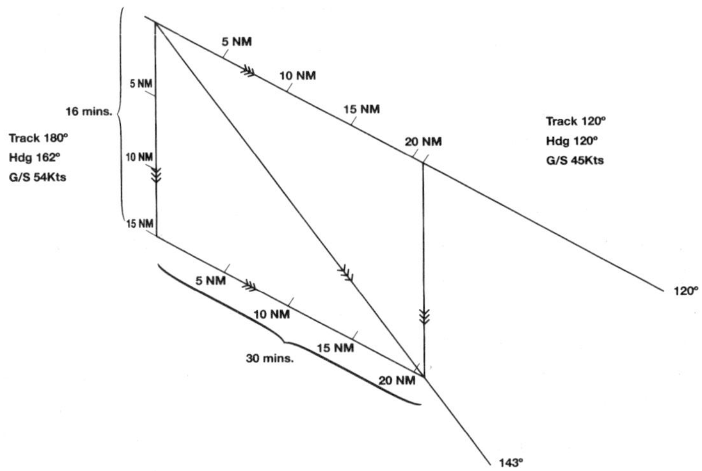

Tracking diagram.

As I took off I could see the weather coming in behind me and it had begun to rain, so everything was the same as usual! I launched myself across the coast. The clouds came racing in behind me. Within minutes the coast disappeared behind a thick black curtain of cloud. I felt a strange sense of contentment; I knew I would make it despite the weather, and I was looking forward to testing my navigational system.

The weather was precisely as forecast: intermittent showers and storms and occasionally very heavy rain. The wind direction and speed were also as forecast, although I often dropped down to 100 feet just to make sure of the speed. I got very wet, but oddly enough I enjoyed the exercise of flying around the storms using the figures I had calculated.

After about four hours, I heard a faint crackle on my radio. It sounded like someone trying to contact me, but the range was too great. My heart lifted a little. I must be close to the mainland! Slowly the weather began to clear, and I was obviously through the worst of it. The cloud base lifted up to 3000 feet, the rain began to ease and visibility improved.

I started to scan the horizon for some indication of land. I had been in the air for five hours, and I knew it must come up soon.

My searching of the horizon was at last greeted by the now familiar thin brown-green line that indicates landfall. I shall never forget that moment. I realised I was going to make it; I was going to get home to see my family and friends. The years of planning, the disastrous start, the 12 000 miles of flying—I had made it.

It was only later that I found out there had been some confusion regarding my track and the weather. At one stage it had been announced on radio and TV that I was missing, and there were grave fears for my safety. My friends, of course, had been desperately trying to keep this information from Jenny and my family. I had flown on, oblivious to all this drama happening on the mainland.

I flew over the mangrove-trimmed coastline and made radio contact with the media plane. Stuart Darbyshire of Regent Air Services was piloting the aircraft and they were at Troughton Island waiting for me, but I planned to head straight to Kalumburu.

Not being exactly sure of my position at that time, I explained to Stuart that I thought I was north of Kalumburu and was tracking in a south-westerly direction down the coast. After about 15 minutes a small settlement came into view with an airstrip next to it. I thought it was probably Kalumburu, and it was. I circled for ten minutes, having been asked by Stuart to wait for them. The Twin Piper Chieftain came into view, and I saw Jenny at the window smiling and waving to me.

Coming across the coast at Kalumburu.
Photographer: Rod Taylor, *The West Australian*

The Chieftain landed, and I came around downwind behind it, but I noticed my engine power slowly dropping, and I pushed the throttle gently forward. Nothing happened, and the engine power continued to fall. I then realised the throttle was full open and the engine power was still dropping—my throttle cable must have snapped somewhere up front. The engine was slowly losing speed, and it seemed I would have to make a glide approach to land.

I bounced onto the end of the runway and came to a fairly quick halt. I was still some 600 metres from the waiting crowd, but with no power I could not taxi. I turned the engine off, took my hands off the controls and my feet off the rudder pedals, lifted my goggles and rested against the back of the seat.

I was home.

I found it hard to believe that I had finally made it. For some reason I felt I just wanted to stay seated in the cockpit and try to get things into perspective. I was shaken out of my reverie by the arrival of a nun on her motorbike. Her veil had almost blown off, and I shall never forget the sight of her smiling face. I was later to find out that S ister Scolastica always travelled by motorbike, and had been at Kalumburu for many years.

'Welcome home, Brian,' she shouted above the noise of the motorbike engine. 'It's good to see you.'

'It's good to see you and to be home,' I replied. Then up roared a truck carrying 20 to 30 Aborigines, obviously painted for a celebration, with Jenny sitting in the front cabin.

I hauled myself stiffly out of the cockpit and stepped onto Australian soil. Jenny jumped out the truck cabin and ran out to me. 'Darling, you've made it. Welcome home,' she said simply. We embraced, both a little misty-eyed.

Neil Donaldson, one of my organising committee, was there to welcome me as well. Then Len Findlay from the West Australian and all the Kalumburu community gathered around the plane. We were all treated to an impromptu corroboree on the airstrip. It really was a marvellous experience.

That night we had dinner with Father Pat, the priest in charge, and the Kalumburu staff. We talked about the history of Kalumburu mission and also Father Pat's desire to have a chapel erected on the coast. Jenny and I decided to be a part of this, and made a donation to the building fund. We were told that we were to be honoured later by being allowed to sleep the night in the sick bay quarters—these quarters had air conditioning! Both of us were feeling very tired. The strain of the past six weeks had taken more of a toll on Jenny than I had contemplated.

We returned to our quarters early, but not before we were treated to another corroboree by the locals. This time they allowed Neil to join in. Stripped to the waist, stamping his feet on the dusty ground in the dull glow of the car headlights, he was hilarious. It was a memorable sight.

That night I slept more comfortably than I had at any other time in the previous five months (although I did wake up on several occasions and rush to the window to check for bad weather). The relief of realising that it didn't matter anymore was almost overwhelming, and on each occasion that I finally realised where I was I would slip back into bed and fall soundly asleep within seconds.

When I awoke the following morning it took a few minutes to work out exactly where I was. The immense relief of realising that it was over flooded through me, and I lay there soaking in the exquisite feeling of contentment—I was home!

My throttle linkage was fixed expertly by an engineer from Alligator Air who flew down from Kununurra, and I took off mid-morning for Broome. I suppose it was knowing that I had arrived home, and was now among family and friends, that made this leg one of the most memorable.

The magnificent rugged landscape, the turquoise ocean and the pearl-white beaches were really quite breathtaking. I flew low over the Mitchell Plateau and the Prince Regent River and finally landed at Broome. Here to meet me were my three children, Michael, Christopher and Susie.

'This is what I call a homecoming,' I thought.

The formalities of customs and immigration were quickly dispensed with, and we all spent a very comfortable night at the Broome Cable Beach Resort. The following morning I departed for Newman. There I was very warmly welcomed at the airport by hundreds of schoolchildren, who said that they had made a special trip out to the airport to watch me arrive. It was here also that I was greeted by my uncle, Ray McNally, whom I had not seen for 30 years.

Next day I took off for Meekatharra via Peak Hill, a small mining town that had requested a visit from me. When I landed at Peak Hill I was presented with a sizeable donation for Legacy, and had tea and sandwiches with the welcoming committee on the airstrip.

On reaching Meekatharra, I was met by my very close friend Tim Keenan. His opening remarks as he shook my hand were, 'God, Eddy, you've taken long enough to get here!' It was great to see him, and we had a very enjoyable night as guests of the local RSL club at their monthly dinner. The following morning I was farewelled by the Leader of the Opposition in the Western Australian Government, Mr Barry MacKinnon, who was in Meekatharra at the time.

atilda
ustralia

TOP LEFT:
A wonderful welcome on arriving at Kalumburu Mission after crossing the Timor Sea.
Photographer Rod Taylor, *The West Australian*

BOTTOM LEFT:
With Jenny at Kalumburu.

ABOVE:
Meeting schoolchildren at Newman, WA.

LEFT:
Signing autographs at Meekatharra, WA.

I had been advised that I was expected to arrive at Langley Park in Perth on Sunday May 13, Mothers Day. The Civil Aviation Authority had given me permission to land there, and promised to erect a temporary control tower. I was beginning to realise that my flight had attracted a lot of attention.

Heavy rain fell throughout the Meekatharra–Cunderdin leg, but I was rather used to this by now and arrived at Cunderdin on time, if very wet. Here again I was met by a large number of the locals, including Mrs Dori Davies, who had welcomed Sir Charles Kingsford Smith some 60 years before on one of his flights across from Victoria. That evening I was officially welcomed by the Shire President.

With Dori Davies at Cunderdin.
Dori greeted Sir Charles Kingsford Smith on one of his flights in WA.
A special photograph!

When I awoke early the following morning, I again found it all hard to believe. I was about to fly the very last leg of my whole adventure. I sat on the edge of my bed, and looked at my torn and somewhat dirty flying jacket, lying crumpled in the corner of my room, my dirty boots on the floor and my flying suit, which was tossed across the chair. All had come a long way with me. I would put them on now for the last time.

I was now very aware of the interest my flight had attracted. I even spoke with Ian McIntosh on the radio program 'Australia All Over' that morning, and found it difficult to contain my excitement: the realisation had hit me that soon I would be landing at Langley Park.

I felt quite emotional during my preflight check that morning. I stood looking at Matilda in the hangar. She was somewhat battered, with a leaking fuel tank, and she was covered in dust and oil, but she only had 80 miles more to go. It looked as though we were going to make it—but only just.

It was overcast as I pushed Matilda out of the hangar and onto the taxi apron, and although thunderstorms were forecast for late in the day, the wind was calm. I was handed a bouquet of flowers by the Ladies Guild of the Uniting Church just before taking off: a Mother's Day present for Jenny.

I left Cunderdin and flew in a south-westerly direction towards Perth. I was to rendezvous over Armadale with Tigers from the Royal Aero Club; they would escort me to Langley Park. As I flew over the Darling Range, the tall buildings of Perth came into view. It was a magnificent sight. Over Armadale my four escorting Tiger Moths picked me up, and we flew in formation down the Canning River and over the Narrows Bridge. Langley Park came into view, with the makeshift runway marked out and with waving and cheering people lining the side. My sense of fulfilment was almost overwhelming.

I came around on final for the last time. Six months previously I had brought Matilda around in the same way. At that time I had had a fleeting thought about my arrival home, and now here it was—I'd done it. The mission was accomplished.

Arrival home:
Langley Park, May 13, 1990.

That final day, May 13, was memorable for many reasons: the 16 000 people there to welcome me home; the 5th Military District band playing Waltzing Matilda; the address by the Premier of Western Australia, Dr Lawrence; the telegrams from the Prime Minister, Bob Hawke, and the Leader of the Opposition, Dr John Hewson, and the lunch at Legacy

House. That wonderful feeling of being among family and friends again: I shall never forget it.

I had had three aims in undertaking the flight. I had done it to honour the memory of my father; to repay Legacy in a small way, and to experience the trials and tribulations and the sheer thrill of flying the same route, and in the same manner, as Australia's magnificent aviation pioneers.

I achieved all three.

With Jenny at Langley

Epilogue

In a strange kind of way, certain things will never be the same. Soon after arriving home I was at a social gathering, and I was asked quite intensely by a friend, 'Were there any spiritual moments on your trip?'

The question took me a little by surprise, not because I hadn't thought about it, but perhaps because I hadn't tried to put any of this in words before. I proceeded to explain, in a very clumsy fashion, the doubts I had experienced during certain parts of the flight—during times of difficult navigation and bad weather. It didn't make sense, and I felt a little embarrassed.

The question made me start reflecting on my trip. I suppose there were a number of times when I felt an odd kind of spiritual presence. There were times when I felt distinctly uncomfortable before taking off—perhaps 'scared' is a more accurate description. Whatever the case, getting myself to climb into the cockpit required a lot of effort.

This apprehension seemed to come upon me after I had been delayed on the ground for a few days, due either to weather or clearance problems. I suppose the delays gave me more time to think about the difficulties of the route segments in front of me, the occasionally desperate nature of dead-reckoning navigation, and the possible mechanical problems with Matilda, which were always lurking. At these times, I would feel decidedly uncomfortable and would have to force myself to adopt an intensely single-minded attitude. I would literally have to force myself to climb into the cockpit.

Once I was in the air, and after some time had passed, this feeling of apprehension would slowly ebb away. On one or two occasions, I remember, my eyes would fill momentarily with tears. I would feel as though I were not alone; it was all very odd.

Flying is a fairly pragmatic activity, with ability almost entirely related to regularity of cockpit occupation. Cockpit drills become second nature, and instrument flying is an activity where most decisions are made for you.

However, there are times when the solo pilot experiences the unique joy and thrill of being in the air, up among the clouds, beyond the constraints of earthly life. It is an exquisite feeling of unexplained joy, a feeling one knows will be short-lived, but never forgotten.

If there is a spiritual aspect to flying maybe it is that. But maybe it's sometimes the thrill of being a man on a mission.

GLOSSARY

NOTE: Flying distances have been given in feet and nautical miles in this book, because this is the normal convention for flying. All other distances are in metres and kilometres.

To convert nautical miles to kilometres, multiply the distance by 1.852. For example, a leg of 400 miles x 1.852 = 740 km.

To convert feet to metres, divide by 3.28. For example, flying at 1600 feet = flying at 487.8 metres above the ground.

ADF: Automatic Direction Finder, a navigational instrument

aileron: device on trailing edge of wing used to bank the aircraft

altimeter: instrument to show height of aircraft above ground

ATC: Air Traffic Control

banking: turning the aircraft

biplane: plane with one wing above the other

chocked: wheel locked by wedge-shaped blocks

compass swing: adjusting compass in the aircraft to allow for deviations due to effects of air frame and engine

crankshaft: rotating shaft within engine to which pistons are attached

crosswind: wind blowing across runway

cruise: maintain constant height and speed

dead-reckoning: using only compass and maps to find location

downwind: flying parallel to runway, opposite to landing direction

drift: amount plane deviates from track due to wind

elevator: control surface on plane used to make aircraft climb or descend

final: last phase of flight before touching down

frequency: radio tuning

fuselage: main section of aircraft containing cockpit

heading: direction of aircraft in relation to compass bearing

Kangaroo Route: route flown from England to Australia in Great Australian Air Race of 1919

leading edge: front part of wing

line up: lining up the aeroplane on the runway ready for take-off

overfly: to fly over airfield at a designated height

RAE: Royal Aerospace Establishment

rudder: control surface used to make plane move from its straight course

spar: main structural part of wing

tachometer: instrument indicating revolutions of engine

tail-dragger: aircraft where tail rests on the ground

tailplane: rudder and elevator section of the aircraft

throttle: instrument to increase or decrease engine power

tracking: direction over the ground the aircraft is travelling in

transponder: instrument to detect aircraft on radar

tricycle undercarriage: three wheels under plane, including a nose wheel

VHF: very high frequency

visual fix: visual reference point for navigation

AWARDS FOR THE MATILDA MISSION

The following awards were made to honour the achievements of the Matilda Mission, which included the raising of $55 000 for Legacy.

On 13 May 1990, Life Membership of Torchbearers for Legacy.

LEFT: **On 4 June 1990, the Epic Achievement Award 1990, from the West Australia Week Council. The award was presented by the Governor of Western Australia, His Excellency Sir Francis Theodore Page Burt AC 1988 KCMG.**

RIGHT: **On 26 January 1991, the Australian Achiever Award from the Australia Day Council, presented by The Honourable Robert Lee Hawke AC 1979.**

LEFT: **On 24 September 1992, at WA Government House, an Order of Australia Medal on the Queens Birthday Honours List for 1992, for Services to Legacy and to Physiotherapy. The award was made by the Lieutenant-Governor and Administrator of WA, His Excellency The Hon. Chief Justice David Kingsley Malcolm AC 1992.**

Photograph courtesy of *The West Australian*.

ABOUT THE AUTHOR

Brian Edwards was born in Perth in 1941. He was educated at Wesley College, South Perth, and later at the University of Western Australia, where he obtained a degree in Science. He also holds a degree in Applied Science from the Curtin University of Technology. Brian is a Fellow of the Australian College of Physiotherapists, and has lectured both nationally and internationally in this field.

Brian is married to Jenny, and they have three children, Michael, Christopher and Susie.

Flying is a hobby of Brian's.